Probability and Stochastic Processes: A Friendly Introduction for Electrical and Computer Engineers

Probability and Stochastic Processes: A Friendly Introduction for Electrical and Computer Engineers

Editor

Giorgos Michel

scitus academics

Probability and Stochastic Processes: A Friendly Introduction for Electrical and Computer Engineers
Edited by **Giorgos Michel**

ISBN: 978-1-68117-452-5
Library of Congress Control Number: 2016937611

www.scitusacademics.com
Box No. 4766, 616 Corporate Way,
Suite 2, Valley Cottage,
NY 10989

Printed in United States of America on Acid Free Paper

Preface

In probability theory, a stochastic process, or often random process, is a collection of random variables representing the evolution of some system of random values over time. This is the probabilistic counterpart to a deterministic process (or deterministic system). Instead of describing a process which can only evolve in one way (as in the case, for example, of solutions of an ordinary differential equation), in a stochastic, or random process, there is some indeterminacy: even if the initial condition is known, there are several directions in which the process may evolve. Classic examples of the stochastic process are guessing the length of a queue at a stated time given the random distribution over time of a number of people or objects entering and leaving the queue and guessing the amount of water in a reservoir based on the random distribution of rainfall and water usage. Stochastic processes were first studied rigorously in the late 19th century to aid in understanding financial markets and Brownian motion. Probability and Stochastic Processes: A Friendly Introduction for Electrical and Computer Engineers covers characterization, structural properties, inference and control of stochastic processes. It is concerned with concepts and techniques, and is oriented towards a broad spectrum of mathematical, scientific and engineering interests.

Table of Contents

Chapter 1

Numerical Approximation of Fractal Dimension of Gaussian Stochastic Processes

Freddy H. Marin Sanchez[1], William Eduardo Alfonso[2]

Basic Science Department, EAFIT University, Carrera 49 # 7 Sur-50, Medellín, Colombia

ABSTRACT

In this paper we propose a numerical method to estimate the fractal dimension of stationary Gaussian stochastic processes using the random Euler numerical scheme and based on an analytical formulation of the fractal dimension for filtered stochastic signals. The discretization of continuous time processes through this random scheme allows us to find, numerically, the expected value, variance and correlation functions at any point of time. This alternative method for estimating the fractal dimension is easy to implement and requires no sophisticated routines. We use simulated data sets for stationary processes of the type Random Ornstein Uhlenbeck to graphically illustrate the results and compare them with those obtained whit the box counting theorem.

KEYWORDS

Stationary Gaussian Stochastic Processes, Fractal Dimension, Random Euler Numerical Scheme

1. INTRODUCTION

A common practice when trying to measure irregular shapes, such as the perimeter of an island or the length of a coastline, is to use Euclidean geometry, ignoring the fact that these shapes do not correspond to the ones of ideal objects, such as polygons and circles, whose dimension is an integer. In contrast, many forms observed in nature have a special feature that makes their resulting measurement depend on the measuring scale, such that the lower the scale, the higher the value of the measurement, to the point where it becomes infinite or indeterminate. The dimension of irregular shapes is a non-integer dimension known as the Hausdorff-Besicovitch dimension. In the early twentieth century, Lewis Fry Richardson found a relationship that allows determining the value of a constant that indicates the degree of roughness of a coast or a geographical border, and used it to calculate the border length between several countries [1]. Based on Richardson's findings, Benoit Mandelbrot [2] found that such irregular shapes have the property of statistical self-similarity, which states that any portion of an irregular shape can be considered a fully scaled image of the entire shape. Later, Mandelbrot coined this property as the fractal dimension to describe sets in $\mathbb{R}^d$. There are many studies in the literature in which maps, border areas and natural and physical phenomena are modeled and simulated with fractal features. In [3] for example, a proposal to simulate real maps using the fractal dimension of Gaussian stochastic processes is presented and some reports modeling terrestrial surfaces, irregular oceanic coasts and rivers, can be found in [4] -[9] .

There are several approaches to estimating of the fractal dimension. In fact in [10] different types of fractals that appear in scientific research are examined, application in nonlinear dynamical systems is discussed, and an exahustive review of numerical methods for their estimation is done.

[11] proposed another method for estimating the fractal dimension of subset of $\mathbb{R}^d$, as well as an additional correction to address the problem of resolution, allowing obtaining not only the fractal dimension, but also a confidence interval for this estimation, thus improving the box counting method. They also show some experimental results based on simulated data and real data from the coastline of Norway. The techniques for the measurement of fractal dimension can be divided into two groups: those based on vectors, which include structured algorithms, and those based on matrices, for the analysis of image systems that include the distance transformation method reported in [12] . Moreover, [13] introduces the roughness method in order to determine the fractal dimension of the Sierpinski carpet, triangle and tetrahedron, the standard Cantor set and the Menger sponge. Meanwhile, in the study of stochastic signals filtering by [14] is obtained the analytical formulation of the fractal dimension to investigate the effects of a linear filter and the regularity in a particular mono-fractal stochastic process, based on an explicit, continuous

formulation, linking the spectral properties of a given filter and the fractal properties of a given process, knowing the function of probability density and output, departing from the correlation coefficient of the filter. With regard to the arguments established by [14] , the aim of our study is to estimate, numerically, the fractal dimension of ergodic stationary Gaussian stochastic processes using the random Euler scheme.

This paper is organized as follows. Section 2 presents the Euler random scheme and builds the numerical solution, the expected value and the crossed expected value for a more general stochastic processe that allows calculating the variance and correlation of the solution of problems of random initial value; the definition of fractal dimension and the box counting theorem are described in Section 3. In Section 4 we present a numericalanalytical formulation for calculating length and fractal dimension for stationary Gaussian stochastic processes. A method to calculate the length and fractal dimension of stationary Gaussian stochastic processes that include processes such as random Ornstein Uhlenbeck is presented in Section 5. Section 6 illustrates graphically, some experimental numerical results associated with the calculation of the fractal dimension of ergodic stationary Gaussian stochastic processes, compared with the results obtained from using the box count theorem.

2. ON THE RANDOM EULER NUMERICAL SCHEME

Consider the initial random value problem of the form1,

$$\begin{aligned} &\dot{X}(t) = f(X(t),t), \quad t \in T = [t_0, t_1] \\ &X(t_0) = X_0 \end{aligned} \tag{1}$$

where X_0 is a random variable and therefore $X(t)$ as $f(X(t),t)$ are both unknown stochastic processes defined on the same probability space $(\Omega, \mathcal{F}, \mathbb{P})$. The aim of this section is to use the Euler scheme to construct a numerical solution, the expected value and the crossed expected value to random initial value problem (1).

Note that the discretized solution of Equation (1) can be expressed in integral form as

$$X_{t_n} - X_{t_{n-1}} = \int_{t_{n-1}}^{t_n} f(X(s), S)\,\mathrm{d}s.$$

By using the approximation $\int_{t_{n-1}}^{t_n} f(S, X(s))\,\mathrm{d}s \approx f(t_{n-1}, X(t_{n-1}))(t_n - t_{n-1})$, the numerical scheme

$$\begin{cases} X_{t_n} = X_{t_{n-1}} + hf\left(X_{t_{n-1}}, t_{n-1}\right) & n \geq 1, \\ X_{t_0} = X_0 \end{cases} \tag{2}$$

is obtained, where X_{t_n}, $f\left(X_{t_{n-1}}, t_{n-1}\right)$ are second order random variables (2-r.v.) and $h = t_n - t_{n-1}$.

In particular, consider a (2-r.v.) X_0 and a stochastic process $b(t)$ second order (2-s.p.) m.s. integrable defined on the same probability space $(\Omega, \mathcal{F}, \mathbb{P})$ such that $b(t)$ is independent of X_0 for each t lying in the interval $T = [t_0, t_1]$. Let $a(t)$ a continuous deterministic function defined on T. The idea is to solve numerically initial random value problems given by the expression

$$\begin{aligned} &\dot{X}(t) = a(t)X(t) + b(t), \quad t \in T = [t_0, t_1] \\ &X(t_0) = X_0. \end{aligned} \tag{3}$$

Note that the Euler numerical scheme for problem (3)2 takes the form

$$\begin{aligned} X_{t_n} &= X_{t_{n-1}} + h\left[a(t_{n-1})X_{t_{n-1}} + b(t_{n-1})\right], \\ &= \left(1 + ha(t_{n-1})\right)X_{t_{n-1}} + hb(t_{n-1}) \quad n \geq 1. \end{aligned} \tag{4}$$

It is easy to show using the principle of mathematical induction, (see [19] and [20] pp. 58-59), that X_{t_n} and $X_{t_n}X_{t_m}$ can be expressed as

$$X_{t_n} = X_{t_0}\prod_{i=0}^{n-1} f_i + h\sum_{i=0}^{n-1}\left(\prod_{j=i+1}^{n-1} f_j\right)b(t_i); \qquad n \geq 1. \tag{5}$$

$$\begin{aligned} X_{t_n}X_{t_m} = {} & X_{t_0}^2\left(\prod_{i=0}^{n-1} f_i\right)\left(\prod_{j=0}^{m-1} f_j\right) + h^2\sum_{i=0}^{n-1}\sum_{k=0}^{m-1}\left(\prod_{j=i+1}^{n-1} f_j\right)\left(\prod_{l=k+1}^{m-1} f_l\right)b(t_i)b(t_k) \\ & + hX_{t_0}\left[\left(\prod_{i=0}^{n-1} f_i\right)\sum_{k=0}^{m-1}\left(\prod_{l=k+1}^{m-1} f_l\right)b(t_k) + \left(\prod_{k=0}^{m-1} f_k\right)\sum_{i=0}^{n-1}\left(\prod_{j=k+1}^{n-1} f_j\right)b(t_i)\right], \end{aligned} \tag{6}$$

where $f_i = 1 + ha(t_i)$. From Equations (5), (6) and the results reported in [19] , it follows that the expected values $E\{X_{t_n}\}$ and $E\{X_{t_n}X_{t_m}\}$ takes the form

$$E\{X_{t_n}\} = E\{X_{t_0}\}\left(\prod_{i=0}^{n-1} f_i\right) + h\sum_{i=0}^{n-1}\left(\prod_{j=i+1}^{n-1} f_j\right)E\{b(t_i)\}, \tag{7}$$

$$E\left\{X_{t_n}X_{t_m}\right\}=E\left\{X_{t_0}^2\right\}\left(\prod_{i=0}^{n-1}f_i\right)\left(\prod_{i=0}^{n-1}f_k\right)+h^2\sum_{i=0}^{n-1}\sum_{k=0}^{m-1}\left(\prod_{j=i+1}^{n-1}f_j\right)\left(\prod_{l=k+1}^{m-1}f_i\right)E\left\{b(t_i)b(t_k)\right\}$$
$$+hE\left\{X_{t_0}\right\}\left[\left(\prod_{i=0}^{n-1}f_i\right)\sum_{k=0}^{m-1}\left(\prod_{l=k+1}^{m-1}f_l\right)E\left\{b(t_k)\right\}+\left(\prod_{k=0}^{m-1}f_k\right)\sum_{i=0}^{n-1}\left(\prod_{l=k+1}^{n-1}f_j\right)E\left\{b(t_i)\right\}\right]. \quad (8)$$

3. FRACTAL DIMENSION AND THE BOX COUNTING THEOREM

The fractal dimension is an objective way to compare fractals geometrically. This section introduces the concept of fractal dimension and presents the box counting theorem.

Let (X,d) denote a complete metric space. Let $A\in H(X)$ be a nonempty compact subset of X. Let $B(x,\epsilon)$ denote the closed ball of radius $\epsilon>0$ and center at a point $x\in X$. We wish to define an integer, $N(A,\epsilon)$, to be the least number of closed balls of radius ϵ needed to cover the set A.

That is, $N(A,\epsilon)=M$: smallest positive integer M such that $A\subset\bigcup_{n=1}^{M}B(x_n,\epsilon)$, for some set of distinct points $\{x_n:n=1,2,\cdots,M\}\subset X$. In other words, surround every point $x\in A$ by an open ball of radius $\epsilon>0$ to provide a cover of A by open sets. Because A is compact this cover possesses a finite subcover, consisting of an integer number, say $\hat{M}$, of open balls. By taking the closure of each ball, thus obtained a cover consisting of M closed balls. Let C denote the set of covers of A by at most $\hat{M}$ closed balls of radius ϵ. Then C contains at least one element.

Let $f:C\rightarrow\{1,2,3,\cdots,M\}$ defined by $f(c)=$ number of balls in the cover $c\in C$. Then $\{f(c):c\in C\}$ is a finite set of positive integers. It follows that it contains a least integer, $N(A,\epsilon)$.

The intuitive idea behind fractal dimension is that a set A has fractal dimension D if: $N(A,\epsilon)\approx C\epsilon^{-D}$ for some positive constant C 3.

Accordingly,

$$D \approx \frac{\ln N(A,\epsilon) - \ln C}{\ln(1/\epsilon)}.$$

Now notice that the term $\frac{\ln C}{\ln(1/\epsilon)}$ approaches 0 as $\epsilon \to 0$. This leads us to the following definition.

Definition 3.1. Let $A \in H(X)$ where (X,d) is a metric space. For each $\epsilon > 0$, let $N(A,\epsilon)$ denote the smallest number of closed balls of radius $\epsilon > 0$ needed to cover A. If

$$D = \lim_{\epsilon \to 0} \left\{ \frac{\ln N(A,\epsilon)}{\ln(1/\epsilon)} \right\}$$

exists, then D is called the fractal dimension A. Also used the notation $D = D_A$ and will say " A has fractal dimension D ".

Theorem 3.2. The Box Counting Theorem. Let $A \in H(\mathbb{R}^m)$, where the Euclidean metric is used. Cover $\mathbb{R}^m$ by closed square boxes of side length $(1/2^n)$. Let $N_n(A)$ denote the number of boxes of side length $(1/2^n)$ which intersect the attractor. If

$$D = \lim_{n \to \infty} \left\{ \frac{\ln(N_n(A))}{\ln(2^n)} \right\},$$

then A has fractal dimension D. The proof of this theorem and other important definitions can be found at [21] .

4. ANALYTICAL FORMULATION OF THE FRACTAL DIMENSION FOR STOCHASTIC PROCESSES

An analytical expression for the fractal dimension of filtered stochastic signals with the purpose of studying the effects of a linear filter and the regularity of a given stochastic mono-fractal process is introduced in [14] . They present some arguments using the Euclidean length of a deterministic curve $x(t)$ with limited duration T, as starting point to estimate the length and fractal dimension of such stochastic signals.

However, stochastic processes can not be described by deterministic continuous curves and irregularities are present even when examined with higher resolutions. This leads to the paradoxical conclusion that the lengths of such signals are not finite. To overcome the fact that stochastic processes lack of a finite longitude, [2] assumes that the longitude L_x of a fractal curve depends on the resolution and the length of the measurement instrument, λ. [1] proves that measuring $L_x(\lambda)$ as a function of λ allows to determine the fractal dimension i.e.:

$$L_x(\lambda) \approx K_x \lambda^{1-D_x} \tag{9}$$

where D_x is the fractal dimension of the curve and K_x is a constant.

Thus, the length L_x can be written as a function of λ given by

$$L_x(\lambda) \approx T\overline{\lambda(h)} \tag{10}$$

where $\overline{\lambda(h)} = \frac{1}{T}\int_{-T/2}^{T/2} \lambda(t,h)\,dt$ and where the elementary length is given by $\lambda(t,h) \approx (1/h)\sqrt{h^2 + \Delta(t,h)^2}$.

This definition of length for continuous curves can be extended to ergodic stationary stochastic processes by replacing the arithmetic mean by the mathematical expectation $E\{.\}$, i.e.

$$L_x(\lambda) = TE[\lambda]$$

where $\lambda = (1/h)\sqrt{h^2 + \Delta^2}$.

Note that $L_x(h) = \frac{T}{h} E\left\{\sqrt{h^2 + \Delta^2}\right\}$ only is solved numerically. In particular, for a high frequency sampling it can be assumed that $h \ll \Delta$ to obtain an analytical expression given by [22]

$$L_x(h) \approx \frac{T}{h} E\{|\Delta|\}. \tag{11}$$

Length and Fractal Dimension for Gaussian Stochastic Processes

Following the results reported in [14] , we will make a brief extension that also requires a prior information such as the probability density function of the process and the variance and correlation functions.

Consider a stochastic process $\{X_t\}_{t\geq 0}$ with values in $\mathbb{R}$ defined on a probability space $(\Omega, \mathcal{F}, \mathbb{P})$. The average length of the process can be estimated by the expression

$$L_x(t) \approx \frac{T}{h} E\{|x_1(t) - x_2(t)|\} \tag{12}$$

with $x_1(t) = X_t, x_2(t) = X_{t-h}; t, t-h \in [0,T]$. Thus, the expected value of the absolute value of the difference between $x_1(t)$ and $x_2(t)$ takes the form

$$E\{|x_1(t) - x_2(t)|\} = \iint |x_1(t) - x_2(t)|\, p_{x_1,x_2}(x_1(t), x_2(t))\, \mathrm{d}x_1 \mathrm{d}x_2, \tag{13}$$

where, $p_{x_1,x_2}(x_1(t), x_2(t))$ denotes the joint probability density function.

Now suppose that the stochastic process $\{X_t\}_{t\geq 0}$ is a stationary Gaussian process. Consequently, $x_1(t)$ and $x_2(t)$ are also Gaussian stationary and therefore the joint probability density can be written as

$$p_{x_1,x_2}(x_1(t), x_2(t)) = \frac{\left(1-\rho_x^2(t)\right)^{1/2}}{2\pi\sigma_1(t)\sigma_2(t)} \mathrm{e}^{\left(-1/\left(2\left(1-\rho_x^2(t)\right)\right)\right)\left(x_1^2(t)/\sigma_1^2(t) + x_2^2(t)/\sigma_2^2(t) - 2\rho_x(t)x_1(t)x_2(t)/\sigma_1(t)\sigma_2(t)\right)}, \tag{14}$$

where $\sigma_1^2(t)$ and $\sigma_2^2(t)$ are the variances of $x_1(t)$ and $x_2(t)$ respectively, and $\rho_x(t)$ is the correlation function defined by $\rho_x(t) = E\{x_1(t)x_2(t)\} / \sqrt{E\{x_1^2(t)\}E\{x_2^2(t)\}}$.

The probability density of the increment $\Delta = x_1(t) - x_2(t)$ is

$$p_\Delta(\Delta) = \int_{-\infty}^{+\infty} p_{x_1 x_2}(x_1(t), x_1(t) + \Delta)\, \mathrm{d}x_1. \tag{15}$$

Analogously to [14] , the probability density of the increment follows a normal distribution $N\left(0, \tilde{\sigma}(t)^2\right)$:

$$p_\Delta(\Delta) = \frac{1}{\sqrt{2\pi\tilde{\sigma}(t)}} e^{-\Delta^2/2\tilde{\sigma}(t)^2}, \tag{16}$$

with $\tilde{\sigma}(t) = \sqrt{\sigma_1^2(t) + \sigma_2^2(t) - 2\rho_x(t)\sigma_1(t)\sigma_2(t)}$. As $x_1(t)$ and $x_2(t)$ belong to the same process $\{X_t\}_{t\geq 0}$ then $\sigma_1(t) = \sigma_2(t) = \sigma_x(t)$ and the standard deviation is $\tilde{\sigma}(t) = \sigma_x(t)\sqrt{2(1-\rho_x(t))}$.

The mathematical expectation of the increment becomes

$$E\{|\Delta|\} = 2\int_0^{+\infty} \Delta p_\Delta(\Delta)\,d\Delta = \sqrt{\frac{2}{\pi}}\tilde{\sigma}(t), \tag{17}$$

and thus

$$L_x(t) \approx \frac{2}{\sqrt{\pi}} \frac{T\sigma_x(t)}{h} \sqrt{1-\rho_x(t)}; \quad \forall t \in [0,T]. \tag{18}$$

Using Equation (9), the fractal dimension D_x of the process $\{X_t\}_{t\geq 0}$ can be approximated as

$$D_x(t,h) \approx 1 - \frac{\log(L_x(t))}{\log(h)} + \frac{\log(K)}{\log(h)}. \tag{19}$$

Finally, Equation (18) becomes

$$D_x(t,h) = 2 - \frac{1}{2}\lim_{h\to 0} \frac{\log(1-\rho_x(t))}{\log(h)}. \tag{20}$$

5. NUMERICAL APPROACH TO FRACTAL DIMENSION FOR STATIONARY GAUSSIAN PROCESSES

In this section we present the numerical construction of the fractal dimension for stationary Gaussian stochastic processes including random Ornstein Uhlenbeck process, using the results of Section 4, the random Euler numerical scheme and the approximation of correlation function using the calculations for $E\{X_{t_n}\}$ and $E\{X_{t_n}X_{t_m}\}$ described in Section 2.

5.1. Calculation of Fractal Dimension for the Random Ornstein Uhlenbeck Process

Consider the random initial value problem given by

$$\dot{X}(t) = -\alpha X(t) + \sigma W(t);\ t \in [0,T] \tag{21}$$
$$X(0) = X_0$$

where $\alpha > 0$, $\sigma > 0$, and $W(t)$ is a Gaussian white noise independent of X_0 for each t lying in the interval $[0,T]$ (defined as in [16] p. 121) and $X_0 \sim N\left(0, \frac{\sigma^2}{2\alpha}\right)$. Note that this process is stationary Gaussian and ergodic, and can be considered as a modification of the random walk in continuous time and analog to the order 1 autoregressive process, AR (1) with discrete time, that broadly describes the velocity of a Brownian mass particle under the influence of friction [23] .

Under random Euler numerical scheme (5), the Equation (21) can be expressed as

$$X_{t_n} = \prod_{i=0}^{n-1}(1-\alpha h) X_0 + \sigma h \sum_{i=0}^{n-1}\left(\prod_{j=i+1}^{n-1}(1-\alpha h)\right) W(t_i);\quad 1 \le n \le N;\ h = \frac{T}{N}. \tag{22}$$

$$X_{t_n} = \beta^n X_0 + \sigma\sqrt{h}\sum_{i=0}^{n-1}\beta^{n-1-i}\xi_i, n \ge 1,\quad \xi_i \sim N(0,1),$$

where $\beta = 1-\alpha h$. By Equation (8) it follows that

$$E\{X_{t_n} X_{t_m}\} = \beta^{n+m} E\{X_0^2\} + \sigma^2 h E\left\{\sum_{i=0}^{n-1}\beta^{n-1-i}\xi_i \sum_{j=0}^{m-1}\beta^{m-1-j}\xi_j\right\}.$$

Trivially note that $E\{X_{t_n}\} = 0$. For $\beta < 1$, $\sum_{i=0}^{n-1}\beta^i = \frac{1-\beta^n}{1-\beta}$; thus

$$E\{X_{t_n} X_{t_m}\} = \beta^{n+m} + \sigma^2 h\left(\frac{1-\beta^{2\min\{n,m\}}}{1-\beta^2}\right).$$

For the case in which $n = N$ and $m = N-1$ it follows that

$$E\left\{X_{t_N}X_{t_{N-1}}\right\}=\beta^{2N-1}+\sigma^2 h\left(\frac{1-\beta^{2(N-1)}}{1-\beta^2}\right)=\beta E\left\{X^2_{t_{N-1}}\right\}.$$

It is easy to see that $\lim_{N\to\infty} E\left\{X^2_{t_N}\right\}=\lim_{N\to\infty} E\left\{X^2_{t_{N-1}}\right\}=\frac{\sigma^2 h}{1-\beta^2}$, and $\lim E\left\{X_{t_N}X_{t_{N-1}}\right\}=\beta\frac{\sigma^2 h}{1-\beta^2}$. In this way,

$$\rho(h)=\frac{\beta\frac{\sigma^2 h}{1-\beta^2}}{\sqrt{\left(\frac{\sigma^2 h}{1-\beta^2}\right)\left(\frac{\sigma^2 h}{1-\beta^2}\right)}}=\beta,$$

which by Equation (20) becomes

$$D_x(h)=2-\frac{1}{2}\lim_{h\to 0}\frac{\log(\alpha h)}{\log(h)}. \tag{23}$$

5.2. Calculation of Fractal Dimension for a Random Mean Reverting Process

Consider the random initial value problem given by

$$\dot{X}(t)=\alpha(\mu-X(t))+\sigma W(t);\ t\in[0,T] \tag{24}$$

$$X(0)=X_0,$$

where $\alpha>0$, $\sigma>0$, $\mu\in\mathbb{R}$, $W(t)$ is a Gaussian white noise independent of X_0 for each t lying in the interval $[0,T]$ (defined as in [16] p. 121) and $X_0\sim N\left(0,\frac{\sigma^2}{2\alpha}\right)$ with $\sigma^2>2\alpha\mu^2$. Note that this process is stationary Gaussian and ergodic. When the initial condition is constant, this process is also known as mean reversion process with additive noise. This kind of process has been widely used in modeling the dynamic behavior of commodity prices and short term interest rates (see [24] and references therein), and the simulating of realistic irregular lattices in [3] .

Equation (24) can be written as

$$\dot{X}(t) = -\alpha X(t) + b(t); \quad b(t) = \alpha\mu + \sigma W(t). \tag{25}$$

Under random Euler numerical scheme, Equation (25) can be expressed as

$$X_{t_n} = \prod_{i=0}^{n-1}(1-\alpha h) X_0 + h\sum_{i=0}^{n-1}\left(\prod_{j=i+1}^{n-1}(1-\alpha h)\right)(\alpha\mu + \sigma W(t_i)); \quad 1 \le n \le N; h = \frac{T}{N}. \tag{26}$$

$$X_{t_n} = \theta\sum_{i=0}^{n-1}\beta^i + \beta^n X_0 + \sigma\sqrt{h}\sum_{i=0}^{n-1}\beta^{n-1-i}\xi_i, n \ge 1, \quad \xi_i \sim N(0,1).$$

Equation (8) implies that for $\theta = \alpha\mu h$ and $\beta = 1 - \alpha h$,

$$E\{X_{t_n} X_{t_m}\} = \theta^2\sum_{i=0}^{n-1}\beta^i\sum_{j=0}^{m-1}\beta^j + \beta^{n+m}E\{X_0^2\} + \theta E\{X_0\}\left(\beta^m\sum_{i=0}^{n-1}\beta^i + \beta^n\sum_{j=0}^{m-1}\beta^j\right) + \sigma^2 h E\left\{\sum_{i=0}^{n-1}\beta^{n-1-i}\xi_i\sum_{j=0}^{m-1}\beta^{m-1-j}\xi_j\right\}.$$

Note that for $\beta < 1$, $\sum_{i=0}^{n-1}\beta^i = \frac{1-\beta^n}{1-\beta}$; in this way, $E\{X_n\} = \theta\left(\frac{1-\beta^n}{1-\beta}\right) + \beta^n E\{X_0\}$.

$$E\{X_{t_n} X_{t_m}\} = \theta^2\left(\frac{1-\beta^n}{1-\beta}\right)\left(\frac{1-\beta^m}{1-\beta}\right) + \beta^{n+m}E\{X_0^2\} + \theta E\{X_0\}\left(\beta^m\left(\frac{1-\beta^n}{1-\beta}\right) + \beta^n\left(\frac{1-\beta^m}{1-\beta}\right)\right) + \sigma^2 h\left(\frac{1-\beta^{2\min\{n,m\}}}{1-\beta^2}\right)$$

For the case in which $n = N$ and $m = N-1$ it follows that $E\{X_{t_N} X_{t_{N-1}}\} = \beta E\{X_{t_{N-1}}^2\} + \theta E\{X_{t_{N-1}}\}$.

It is easy to see that $\lim_{N\to\infty} E\{X_{t_N}\} = \frac{\theta}{1-\beta}$; $\lim_{N\to\infty} E\{X_{t_N}^2\} = \lim_{N\to\infty} E\{X_{t_{N-1}}^2\} = \frac{\theta^2}{(1-\beta)^2} + \frac{\sigma^2 h}{1-\beta^2}$

and

$\lim_{N\to\infty} E\{X_{t_N} X_{t_{N-1}}\} = \beta\frac{\theta^2}{(1-\beta)^2} + \beta\frac{\sigma^2 h}{1-\beta^2} + \frac{\theta^2}{1-\beta}$. In this way,

$$\rho(h) = \frac{\beta\frac{\theta^2}{(1-\beta)^2} + \beta\frac{\sigma^2 h}{1-\beta^2} + \frac{\theta^2}{1-\beta}}{\sqrt{\left(\frac{\theta^2}{(1-\beta)^2} + \frac{\sigma^2 h}{1-\beta^2}\right)\left(\frac{\theta^2}{(1-\beta)^2} + \frac{\sigma^2 h}{1-\beta^2}\right)}} = \frac{\left(\mu^2 + \frac{\sigma^2}{2\alpha - \alpha^2 h}\right)(1-\alpha h) + \alpha\mu^2 h}{\left(\mu^2 + \frac{\sigma^2}{2\alpha - \alpha^2 h}\right)} = 1 - \frac{\alpha\sigma^2 h}{\sigma^2 + \mu^2(2\alpha - \alpha^2 h)}$$

Equation (20) becomes

$$D_x(h) = 2 - \frac{1}{2}\lim_{h\to 0}\left[\frac{\log\left(\frac{\alpha\sigma^2 h}{\sigma^2 + \mu^2\left(2\alpha - \alpha^2 h\right)}\right)}{\log(h)}\right]. \tag{27}$$

6. NUMERICAL RESULTS

This section graphically illustrates some experimental numerical results related to the calculation of the fractal dimension for stationary Gaussian stochastic processes associated with random initial value problems (21) and (24).

Figure 1(a) shows a comparison between the approximate average fractal dimension of 1000 simulated trajectories for Ornstein Uhlenbeck process using the Box Counting theorem and the numerical fractal dimension proposed in Equation (23) for different values of h. In Figure 1(b) the dynamic behavior of the fractal dimension is shown for different values of the parameter α .

Figure 2(a) shows a comparison between the approximate average fractal dimension of 1000 simulated trajectories for mean reverting process using the Box Counting theorem and the numerical fractal dimension proposed in Equation (27) for different values of h. In Figure 2(b) the dynamic behavior of the fractal dimension is shown for different values of the parameter α .

7. CONCLUSIONS

This paper made a slight extension of the approach proposed in [14] for the numerical estimation of the fractal dimension of ergodic stationary Gaussian stochastic processes associated with the random Ornstein Uhlenbeck process. Using random Euler scheme performed numerical construction of the expectation value, variance and correlation functions at any point of time to initial random value problems general. The computational implementation of this alternative method to estimate the fractal dimension is easy and requires no sophisticated routines. It illustrates graphically the results obtained compared with box counting theorem using simulated data sets for stationary processes type random Ornstein Uhlenbeck.

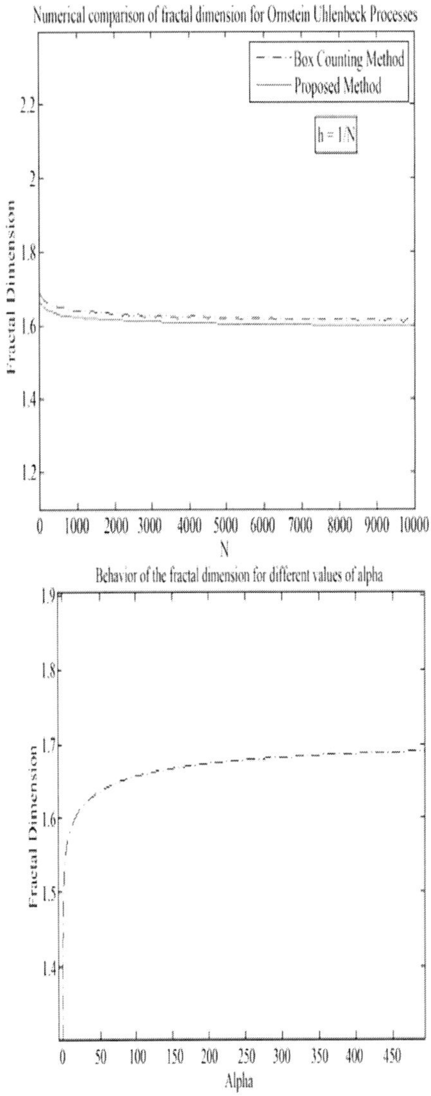

Figure 1. (a) Numerical comparison of fractal dimension for Ornstein Uhlenbeck processes with $T=1$, $\alpha=10$ and $\sigma=20$ for different values of h; (b) Dynamic behavior of the fractal dimension for Ornstein Uhlenbeck processes with $T=1$, $h=1/10^4$ and $\sigma=20$ for $0.001 \le \alpha \le 500$.

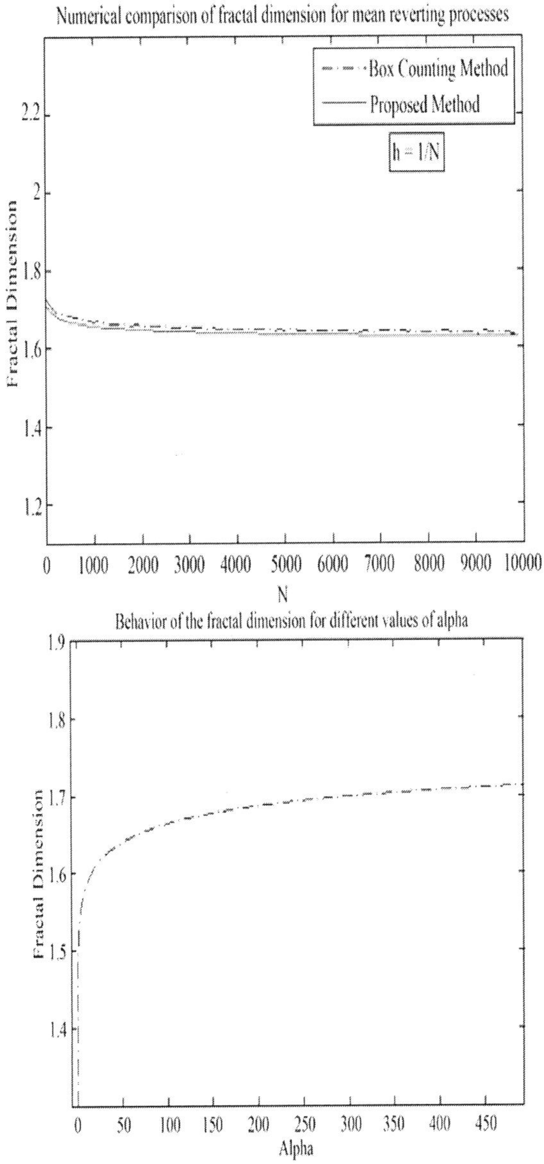

Figure 2. (a) Numerical comparison of fractal dimension for mean reverting processes with $T=1$, $\mu=10$, $\alpha=20$ and $\sigma=250$ for different values of h; (b) Dynamic behavior of the fractal dimension for mean reverting processes with $T=1$, $\mu=10$, $h=1/10^4$ and for $0.001 \le \alpha \le 500$.

A clear convergence of the proposed methods is shown in Figure 1(a) and Figure 2(a). In addition, the results show that the fractal dimension of Ornstein

Uhlenbeck processes is independent of the parameter μ . This independence is maintained in the mean reverting process for the case where $\sigma^2 > 2\alpha\mu^2$. Figure 1(b) and Figure 2(b) reflect the fact that the fractal dimension of this type of process is more sensitive to the parameter α .

REFERENCES

1. Richardson, L. (1961) The Problem of Contiguity: An Appendix of Statistic of Deadly Quarrels. General Systems Year Book, 61, 139-187.

2. Mandelbrot, B. (1967) How Long Is the Coast of Britain Statistical Self-Similarity and Fractional Dimension. Science, 156, 636-638.

3. Duque, J.C., Betancourt, A. and Marin, F. (2014) An Algorithmic Approach for Simulating Realistic Irregular Lattices. In: Thill, J.C. and Dragicevic, S., Eds., GeoComputational Analysis and Modeling of Regional Systems, Springer Heidelberg, in press.

4. Mandelbrot, B. (1975) Stochastic Models for the Earth's Relief, the Shape and the Fractal Dimension of the Coastlines, and the Number-Area Rule for Islands. Proceedings of the National Academy of Sciences, 72, 3825-3828.

5. Adler, R.J. (1978) Some Erratic Patterns Generated by the Planar Wiener Process. Advances in Applied Probability, 10, 22-27.

6. Cullin, W.E. and Datko, M. (1987) The Fractal Geometry of the Soil-Covered Landscape. Earth Surface Processes and Landforms, 12, 369-385.

7. Cullin, W.E. (1988) Dimension and Entropy in the Soil-Covered Landscape. Earth Surface Processes and Landforms, 13, 619-648.

8. Beauvais, A. Montgomery, D. (1996) Influence of Valley Type on the Scaling Properties of River Planforms. Water Resources Research, 32, 1441-1448.

9. Montgomery, K. (1996) Sinuosity and Fractal Dimension of Meandering Rivers. Area, 28, 491-500.

10. Theiler, J. (1990) Estimating Fractal Dimension. JOSA A, 7, 1055-1073.

11. Taylor C. and Taylor, S.J. (1991) Estimating the Dimension of a Fractal. Journal of the Royal Statistical Society. Series B (Methodological), 53, 353-364.

12. Allen, M., Brown, G.J. and Miles, N.J. (1995) Measurement of Boundary Fractal Dimensions: Review of Current Techniques. Powder Technology, 84, 1-14.

13. Blachowski, A. and Ruebenbauer, K. (2009) Roughness Method to Estimate Fractal Dimension. Acta Physica Polonica A, 115, 636-640.

14. Girault, J., Kouame, D. and Ouahabi, A. (2010) Analytical Formulation of the Fractal Dimension of Filtered Stochastic Signals. Signal Processing, 90, 2690-2697.

15. Loeve, M. (1963) Probability Theory. D. Van N. Company, London.

16. Soong, T. (1973) Random Differential Equations in Science and Engineering. Academic Press, New York.

17. Jodar, L., Cortez, J. and Villafuerte, L. (2007) Mean Square Numerical Solution of Random Differential Equations: Facts and Possibilities. Computers and Mathematics with Applications, 53, 1098-1106.

18. Jodar, L. and Villafuerte, L. (2007) Numerical Solution of Random Differential Equations: A Mean Square Approach. Mathematical and Computer Modelling, 45, 757-765.

19. Marin, F. and Laniado, H. (2014) Convergence of Analytical Stochastic Processes in Mean Square. 1-9. http://repository.eafit.edu.co/handle/10784/1523

20. Agarwall, R. (1992) Difference Equations and Inequalities. Marcel Dekker, New York.

21. Barnsley, M. (1988) Fractals Everywhere. Academic Press, London.

22. Le Tavernier, E. (1998) La methode de Higuchi pour la dimension fractal. Signal Processing, 65, 115-128. Uhlenbeck, G. and Ornstein, L. (1930) On the Theory of the Brownian Motion. Physical Review, 36, 823-841.

23. Marin, F. and Palacio, J. (2013) Gaussian Estimation of One-Factor Mean Reversion Processes. Journal of Probability and Statistics, 2013, 10 p.

Chapter 2

A notion of stability in probability of stochastic nonlinear systems

Fakhreddin Abedi[1] Wah June Leong[1] and Sarkhosh Seddighi Chaharborj

Department of Mathematics, Faculty of Science, Universiti Putra Malaysia (UPM)

ABSTRACT

The concepts of stability in probability of nontrivial solutions for stochastic nonlinear systems are analyzed in terms of a control Lyapunov function which is smooth except possibly at the origin. We show under certain hypothesis that the neighborhood of the origin is stable in probability. An illustrating example is provided.

KEYWORDS

stability in probability stochastic nonlinear systems Lyapunov technique

1 INTRODUCTION

In recent years, the stability analysis of stochastic nonlinear systems (SNS) has been one of the most active and important areas in control theory. Although there are many research papers (see, for instance, [1–13]) on the stability of trivial solutions of SNS, but there are very few results on the stability of nontrivial solutions of stochastic systems. It is known that there are many types of stochastic systems all of whose solutions tend to each other although they do not have the trivial solution. These systems are worth being interested in since the solutions with different initial value has similar large-time properties. In this paper, we study the stability in probability of a nontrivial solution of the SNS.

The main results of the present paper is to explore further the stabilization in probability problem for a wider class of SNS than that described in [14]. This paper is also intended to fill in the gap of the previous works by establishing necessary and sufficient conditions for stability in probability of nontrivial solution for the SNS. Our main result, Theorem 3.1, which is an extension of Theorem 3.1 established in [2], asserts that the SNS is stable in probability if and only if it admits a control Lyapunov function (CLF). The main tool used in this paper is a stochastic version of the converse Lyapunov theorem established in Kushner [9].

The paper is organized as follows. In Section 2, we introduce a class of stochastic systems, some basic definitions and properties of stochastic CLF which play an important role in the stability theory. In Section 3, we state and prove the main results of the paper on the stability in probability of nontrivial solution of SNS. Finally, in Section 4, we provide a numerical example to validate our results.

2 STOCHASTIC STABILITY

In this section we first introduce a class of stochastic systems and define the stability in probability of nontrivial solutions of this system. We also focus on the properties of CLF which play an important role to the stability in Section 3.

Let (Ω,F,P) (Ω,F,P) be a complete probability space, and denote by $(w\ t)\ t\geq 0$ (wt)t≥0 a standard $\mathbb{R}^m$-valued Wiener process defined on this space.

Consider the following SNS

$$dx = f(t,x)\,dt + g(t,x)\,dw_t, \tag{1}$$

where $x \in \mathbb{R}^n$ is the system state vector, f: $\mathbb{R}^+ \times \mathbb{R}^n \to \mathbb{R}^n$ and $g : \mathbb{R}^+ \times \mathbb{R}^n \to \mathbb{R}^{n\times m}$ are continuous with $f(t,0) \neq 0,\ g(t,0) \neq 0$, and there exist constants $C_1, C_2 \geq 0$ such that the following conditions hold:

$$|f(t,x) - f(t,y)| + |g(t,x) - g(t,y)| \leq C_1(|x-y|),$$

$$|f(t,x)| + |g(t,x)| \leq C_2(1 + |x|).$$

We note that SNS (1) satisfies a sufficient condition, under which there is some unique global solution $x(t,x_0)$ x(t,x0) to SNS (1) for any given initial value $x(0)=x_0 \in R^n$ x(0)=x0∈Rn (see, for instance, Khasminskii [8]).

Throughout this paper we adopt the following notations:

- For any $x_0 \in R^n$ x0∈Rn, denote by $x(t)=x(t,x_0)$ x(t)=x(t,x0) the solution of SNS (1).
- Denote by $|\cdot|$ |·| the usual Euclidean norm in R^n Rn.
- We use $\Phi_0=\Phi(0,x_0)$ Φ0=Φ(0,x0) to denote the initial value of a Lyapunov function.
- N N denotes the class of nonnegative C^0 C0 functions $\mu:R_+\to R_+$ μ:R+→R+, for which $\int_0^{+\infty}\mu(t)\,dt < +\infty$ and $\lim_{t\to+\infty}\mu(t)=0$ $\lim_{t\to+\infty}\mu(t)=0$ limt→+∞μ(t)=0 hold.
- $S_r=\{x\in R^n;|x|\leq r,r>0\}$ Sr={x∈Rn;|x|≤r,r>0} is a ball.
- We define $\tau\wedge t=\min\{\tau,t\}$ τ∧t=min{τ,t}.

Definition 2.1 A function $\gamma:R_+\to R_+$ γ:R+→R+ is a

1. function if it is continuous, strictly increasing and $\gamma(0)=0$ γ(0)=0, and
2. K_∞ K∞ -function if it is a K-function and also $\gamma(r)\to\infty$ as $r\to\infty$.

We now give the definition of stability in probability for SNS (1) when the origin is not a trivial solution. In this case we study the stability in probability of solutions with respect to the small neighborhood of the origin.

Definition 2.2 We say that S_r Sr, $r>0$ r>0 is stable in probability if for any $c>r$ c>r and any $\epsilon\in(0,1)$ ϵ∈(0,1), there exists $\delta=\delta(\epsilon,c)$ δ=δ(ϵ,c) such that for all $t\geq t_0\geq 0$ t≥t0≥0 and $|x_0|<\delta$ |x0|<δ, the following condition holds:

$$P\{|x(t)| < c\} \geq 1 - \epsilon. \tag{2}$$

In the following we consider the SNS

$$dx = \left(f(t,x) + h(t,x)u\right)dt + g(t,x)\,dw_t, \tag{3}$$

where $x \in \mathbb{R}^n$, $u \in \mathbb{R}^p$, the dynamics $f(t,x)$, $g(t,x)$ and $h : \mathbb{R}^+ \times \mathbb{R}^n \to \mathbb{R}^{n \times p}$, are continuous and Lipschitz functions with $f(t,0)\neq 0$ f(t,0)≠0 and $g(t,0)\neq 0$ g(t,0)≠0.

Theorem 3.1 of this paper is an extension of Theorem 3.1 provided in [2] that guarantees the existence of a $C\infty$ C∞ control law $u=\phi(t,x)$ u=ϕ(t,x) in such a way that Sr Sr satisfy the stability in probability property with respect to the closed-loop system

$$dx = \left(f(t,x) + h(t,x)\phi(t,x)\right) dt + g(t,x)\, dw_t. \tag{4}$$

Denote by **D** the infinitesimal generator of the stochastic process solution of the uncontrolled part of SNS (3), that is, **D** is the second-order differential operator defined for any function Φ in $C^{2,1}(\mathbb{R}^+ \times \mathbb{R}^n, \mathbb{R})$ by

$$\mathbf{D}\Phi(t,x) = \frac{\partial \Phi(t,x)}{\partial t} + \sum_{i=1}^{n} f^i \frac{\partial \Phi(t,x)}{\partial x_i} + \frac{1}{2} \sum_{i,j=1}^{n} \sum_{k=1}^{m} g_k^i g_k^j \frac{\partial^2 \Phi(t,x)}{\partial x_i \, \partial x_j}.$$

For any $z\in(1,\ldots,p)$ z∈(1,…,p), denote by $\mathbf{D}z$ Dz the first-order differential operator defined for any function Φ in $C^{1,1}(\mathbb{R}^+ \times \mathbb{R}^n, \mathbb{R})$ by

$$\mathbf{D}_z \Phi(t,x) = \sum_{i=1}^{n} h_z^i(t,x) \frac{\partial \Phi(t,x)}{\partial x_i}.$$

Define **Y** as the infinitesimal generator for the stochastic process solution of the closed-loop system (4), that is, **Y** is the differential operator defined for any function Φ in $C^{1,1}(\mathbb{R}^+ \times \mathbb{R}^n, \mathbb{R})$ by

$$\mathbf{Y}\Phi(t,x) = \mathbf{D}\Phi(t,x) + \sum_{z=1}^{p} \mathbf{D}_z \Phi(t,x) u_z.$$

In the following we extend the concept of stochastic Lyapunov condition introduced in Definition 2.4 provided in [2] used for stability in probability of SNS (3) at the neighborhood of the origin.

Definition 2.3 SNS (3) admits a CLF if there exists a $C2,1$ C2,1 function $\Phi : \mathbb{R}^+ \times \mathbb{R}^n \to \mathbb{R}^+$, class $K\infty$ K∞ functions $a1$ a1, $a2$ a2, a $C0$ C0 positive definite function $\rho : \mathbb{R}^+ \to \mathbb{R}^+$, and a nonnegative $C0$ C0 function μ of class N N such that for all $(t,x)\in R+\times Rn$ (t,x)∈R+×Rn, the following conditions hold:

$$a_1(|x|) \leq \Phi(t,x) \leq a_2(|x|), \tag{5}$$

$$\mathrm{D}_z\Phi(t,x) = 0 \quad \Rightarrow \quad \mathrm{D}\Phi(t,x) \leq -\rho(\Phi(t,x)) + \mu(t). \tag{6}$$

Remark 2.4 We note that condition (6) obtained in Definition 2.3 can be replaced with the following condition

$$\mathrm{D}\Phi(t,x) + \sum_{z=1}^{p} \mathrm{D}_z\Phi(t,x)u_z \leq -\rho(\Phi(t,x)) + \mu(t).$$

In the next section we will turn the attention to the class of SNS (3) and provide some results related to the stability in probability of this system.

3 MAIN RESULTS

In this section, where the dynamics *f*(*t*,0) f(t,0) and *g*(*t*,0) g(t,0) may be non-vanishing at the origin, we study the stability in probability of nontrivial solutions of SNS (3). In the following theorem that is an extension of Theorem 3.1 established in [2], we derive the necessary and sufficient conditions for stability in probability of nontrivial solutions of SNS (3).

Theorem 3.1 *Consider SNS* (3). *Then the following three statements are equivalent*:

1. *There exists a* C^∞ C∞ *control law* $u=\phi : \mathbb{R}^+ \times \mathbb{R}^n \to \mathbb{R}^m$ *for all* $t\geq 0$ t≥0 *such that* S_r Sr *satisfy the stability in probability property with respect to closed-loop system* (4).

2. *There exists a* C^0 C0 *control law* $u = \phi : \mathbb{R}^+ \times \mathbb{R}^n \to \mathbb{R}^m$ *with* $\phi(t,0)=0$ ϕ(t,0)=0 *for all* $t\geq 0$ t≥0, *being locally Lipschitz in x, in such a way that* S_r Sr *satisfy the stability in probability property with respect to closed-loop system* (4).

3. (*There exists a* $C^{2,1}$ C2,1 *function* $\Phi : \mathbb{R}^+ \times \mathbb{R}^n \to \mathbb{R}^+$, *class* K_∞ *class* K_∞ K∞ *functions* a_1 a1, a_2 a2, *a* C^0 C0 *positive definite function* $\rho:\mathbb{R}^+\to\mathbb{R}^+$ ρ:R+→R+, *and a nonnegative* C^0 C0 *function* μ *of class* N N *such that for all* $(t,x)\in\mathbb{R}^+\times\mathbb{R}^n$ (t,x)∈R+×Rn, *conditions* (5) *and* (6) *hold*.

Proof The implication (i) → (ii) is obvious.

We should establish the implication (ii) → (iii). Suppose that there exists a C^0 C0 control law $u=\phi(t,x)$ u=ϕ(t,x) (as in statement (ii) of the theorem) such that S_r Sr satisfies the stability in probability with respect to the closed-loop system (4). Then, by the converse Lyapunov theorem (Theorem 2) developed by Kushner [9], which proved the existence of a stochastic Lyapunov function in some neighborhood of the origin, there exists a Lyapunov function Φ in $C^{2,1}(\mathbb{R}^+\times\mathbb{R}^n,\mathbb{R}^+)$ C2,1(R+×Rn,R+) and a continuous and positive definite function $k(t,x)$ k(t,x) in such a way that (5) holds and

$$\mathbf{Y}\Phi(t,x) = -k(t,x) \leq 0, \tag{7}$$

where **Y** is the infinitesimal generator for the stochastic process solution of closed-loop system (4). The latter inequality implies (6) with $\rho(\Phi(t,x))=k(t,x)$ ρ(Φ(t,x))=k(t,x), $\mu(t)=0$ μ(t)=0 and $\mathbf{D}_z\Phi(t,x)=0$ DzΦ(t,x)=0. Therefore, Definition 2.3 holds and the Lyapunov function $\Phi(t,x)$ Φ(t,x) is a stochastic CLF for SNS (3).

Next, for (iii) → (i), suppose that there exists a $C^{2,1}$ C2,1 function $\Phi:\mathbb{R}^+\times\mathbb{R}^n\to\mathbb{R}^+$ such that both conditions (5) and (6) are fulfilled with respect to closed-loop system (4). Without any loss of generality, assume

$$\mu(t) > 0 \quad \forall t \geq 0. \tag{8}$$

Notice, by virtue of (5), that

$$\frac{\partial\Phi}{\partial t}(t,0) = 0, \qquad \frac{\partial\Phi}{\partial x}(t,0) = 0. \tag{9}$$

Then using (5) and (6) and by applying the converse result of Theorem 3.1 established in [2], condition (6) in conjunction with (8) and (9) enables us to build, by the standard partition of unity arguments, a C^∞ C∞ control law $u=\phi(t,x)$ u=ϕ(t,x) with $\phi(t,0)=0$ ϕ(t,0)=0 such that

$$\mathbf{Y}\Phi(t,x) = \mathbf{D}\Phi(t,x) + \sum_{z=1}^{p}\mathbf{D}_z\Phi(t,x)u_z \leq -\rho\big(\Phi(t,x)\big) + \mu(t). \tag{10}$$

From (10) and $\Phi(t,x)\geq 0$ Φ(t,x)≥0, we have that $\Phi_t=\Phi(t,x)$ Φt=Φ(t,x) is a supermartingale. Let $\tau=\inf\{t\geq 0;|x(t)|\geq c, c>r\}$ τ=inf{t≥0;|x(t)|≥c,c>r} and

$$\Phi\big(\tau\wedge t, x(\tau\wedge t)\big) = \Phi_0 + \int_0^{\tau\wedge t}\left(\mathbf{D}\Phi(s,x) + \sum_{z=1}^{p}\mathbf{D}_z\Phi(s,x)u_z\right)ds$$
$$+\int_0^{\tau\wedge t}\sum_{i=1}^{n} g^i(s,x)\frac{\partial\Phi(s,x)}{\partial x_i}\,dw_s. \tag{11}$$

With (10) and (11) we have

$$\Phi(\tau \wedge t, x(\tau \wedge t)) = \Phi_0 + \int_0^{\tau \wedge t} \mathbf{Y}\Phi(s,x)\, ds + \int_0^{\tau \wedge t} \sum_{i=1}^{n} g^i(s,x) \frac{\partial \Phi(s,x)}{\partial x_i}\, dw_s. \tag{12}$$

From (12) we obtain

$$E[\Phi(\tau \wedge t, x(\tau \wedge t))] = \Phi_0 + E\left(\int_0^{\tau \wedge t} \mathbf{Y}\Phi(s,x)\, ds\right). \tag{13}$$

Assume that there exists a nonnegative *C* 0 C0 function *μ* of class N N such that

$$\int_0^{+\infty} \mu(t)\, dt \leq K < +\infty \tag{14}$$

for all *K*>0 K>0. Taking into account (10) and (13)-(14) yields

$$\begin{aligned} E[\Phi(\tau \wedge t, x(\tau \wedge t))] &\leq \Phi_0 + E\left(\int_0^{\tau \wedge t} (-\rho(\Phi(s,x)) + \mu(s))\, ds\right) \\ &\leq \Phi_0 + E\left(\int_0^{\tau \wedge t} \mu(s)\, ds\right) \\ &\leq \Phi_0 + E\left(\int_0^{+\infty} \mu(s)\, ds\right) \\ &\leq \Phi_0 + K. \end{aligned} \tag{15}$$

Let for a given *ϵ*∈]0,1[ϵ∈]0,1[there exist *δ*=*δ*(*ϵ*,*c*)∈]0,*c*[δ=δ(ϵ,c)∈]0,c[, *c*>*r* c>r, such that for *x* 0 ∈*S δ* x0∈Sδ and *a* 1 (*c*)>*Kϵ* a1(c)>Kϵ, the following inequality holds

$$\frac{\Phi_0}{\epsilon} \leq a_1(c) - \frac{K}{\epsilon}. \tag{16}$$

We note that |*x*(*τ*∧*t*)|=|*x*(*τ*)|=*c* |x(τ∧t)|=|x(τ)|=c if *τ*≤*t* τ≤t. So,

$$E[\Phi(\tau \wedge t, x(\tau \wedge t))] \geq E(1_{(\tau \leq t)}\Phi(\tau, x(\tau))). \tag{17}$$

Then, by taking into account (5) and (17), we have

$$E\big[\Phi(\tau \wedge t, x(\tau \wedge t))\big] \geq E\big(1_{(\tau \leq t)} a_1\big(|x(\tau)|\big)\big)$$
$$\geq E\big(1_{(\tau \leq t)} a_1(c)\big) = a_1(c) E(1_{(\tau \leq t)})$$
$$= a_1(c) P(\tau \leq t). \tag{18}$$

Thus, from (15)-(16) and (18), we have

$$a_1(c) P(\tau \leq t) \leq \Phi_0 + K \leq \epsilon a_1(c). \tag{19}$$

The latter inequality implies that

$P(\tau \leq t) \leq \epsilon.$ Letting $t \to +\infty$ t→+∞, we obtain

$P(\tau \leq +\infty) \leq \epsilon$. P(τ≤+∞)≤ϵ.

This implies that for given $\epsilon, r \geq 0$ ϵ,r≥0 and $c > r$ c>r, there exists $\delta \in]0, c[$ δ∈]0,c[such that

$$P\big\{|x(t)| < c\big\} \geq 1 - \epsilon \tag{20}$$

for all $t \geq t_0 \geq 0$ t≥t0≥0 and $|x_0| < \delta$ |x0|<δ. Therefore, from (20) and Definition 2.2, S_r Sr is stable in probability with respect to closed-loop system (4). This completes the proof of Theorem 3.1. □

4 APPLICATION

In this section we illustrate our results by designing a numerical example.

Example 4.1 Consider the SNS

$$d\begin{pmatrix} x_1 \\ x_2 \end{pmatrix} = \begin{pmatrix} x_1 + x_2^3 \\ 2 \end{pmatrix} dt + \begin{pmatrix} 0 \\ 1 \end{pmatrix} u\,dt + \begin{pmatrix} x_1 \\ \frac{1}{2} \end{pmatrix} dw_t, \tag{21}$$

where $(w_t)_{t \geq 0}$ (wt)t≥0 is a standard real-valued Wiener process, u is a real-valued measurable control law, $f_1(t, x_1, x_2) = x_1 + x_2^3, f_2(t, x_1, x_2) = 2, h_1(t, x_1, x_2) = 0, h_2(t, x_1, x_2) = 1, g_1(t, x_1, x_2) = x_1,$ and $g_2(t, x_1, x_2) = \frac{1}{2}.$ Define the Lyapunov function in the form

$\Phi(t, x_1, x_2) = 2x_1^2 \exp(2t) + (x_2 + x_1 \exp(t))^4$. Φ(t,x1,x2)=2x12exp(2t)+(x2+x1exp(t))4.

A simple calculation shows that

$$\frac{\partial \Phi(t,x_1,x_2)}{\partial x_2} = 0 \quad \Leftrightarrow \quad x_2 = -x_1 \exp(t). \tag{22}$$

From (22) we
$\Phi(t,x_1,x_2) = 2x_1^2 \exp(2t)$, and so $\frac{\partial \Phi}{\partial t} = 4x_1^2 \exp(2t)$, $\frac{\partial \Phi}{\partial x_1} = 4x_1 \exp(2t)$, $\frac{\partial^2 \Phi}{\partial x_1^2} = 4\exp(2t)$, and $\frac{\partial \Phi}{\partial x_2} = \frac{\partial^2 \Phi}{\partial x_2^2} = \frac{\partial^2 \Phi}{\partial x_1 \partial x_2} = \frac{\partial^2 \Phi}{\partial x_2 \partial x_1} = 0.$

Thus, we get

$$\mathbf{D}_z \Phi(t,x_1,x_2) = \sum_{i=1}^{n} h_z^i(t,x) \frac{\partial \Phi(t,x)}{\partial x_i} = 0,$$

and

$$\begin{aligned}
\mathbf{D}\Phi(t,x_1,x_2) &= \frac{\partial \Phi(t,x)}{\partial t} + \sum_{i=1}^{n} f^i(t,x) \frac{\partial \Phi(t,x)}{\partial x_i} \\
&\quad + \frac{1}{2} \sum_{i,j=1}^{n} \sum_{k=1}^{m} g_k^i g_k^j \frac{\partial^2 \Phi(t,x)}{\partial x_i \partial x_j} \\
&= 4x_1^2 \exp(2t) + \left(x_1 + x_2^3\right) 2x_1 \exp(2t) + 4x_1^2 \exp(2t) \\
&= 6x_1^2 \exp(2t) + \left(x_1 - x_1^3 \exp(3t)\right) 4x_1 \exp(2t) \\
&= 10x_1^2 \exp(2t) - 4x_1^4 \exp(5t).
\end{aligned}$$

Therefore,

$$\begin{aligned}
\mathbf{Y}\Phi(t,x_1,x_2) &= \mathbf{D}\Phi(t,x) + \sum_{z=1}^{p} \mathbf{D}_z \Phi(t,x) u_z \\
&= 10x_1^2 \exp(2t) - 4x_1^4 \exp(5t) \\
&= 5\Phi(t,x_1,x_2) - \exp(t)\Phi^2(t,x_1,x_2) \\
&\leq -\frac{1}{2}\Phi^2(t,x_1,x_2) + 5\exp(-t).
\end{aligned}$$

The latter inequality implies that both (5) and (6) are fulfilled with *ρ*(*s*)=12 *s* 2 ρ(s)=12s2, *μ*(*t*)=5exp(−*t*) μ(t)=5exp(−t), *a* 1 (*s*)=12 *s* 2 a1(s)=12s2 and *a* 2 (*s*)=6*s* 2 a2(s)=6s2, and thus, by Theorem 3.1, there exists a *C* ∞ C∞ - feedback law *ϕ*(*t*,*x* 1 ,*x* 2) ϕ(t,x1,x2) with *ϕ*(*t*,0,0)=0 ϕ(t,0,0)=0 such that *S r* Sr is stable in probability with respect to the resulting closed-loop system deduced from (21).

5 CONCLUSIONS

Many authors have paid attention to the subject of stability in probability of trivial solutions of stochastic systems in the past years. On the other hand, it is known that there are many types of stochastic systems all of whose solutions tend to each other although they do not have the trivial solution. In this paper, we have established the stability in probability of nontrivial solutions for SNS (3). We have derived a stochastic version of CLF and provided the necessary and sufficient conditions for stability in probability of a nontrivial solution for SNS (3).

ACKNOWLEDGEMENTS

The authors thank the referees for valuable comments and suggestions which improved the presentation of this manuscript.

AUTHORS' CONTRIBUTIONS

All authors read and approved the final manuscript.

REFERENCES

1. Abedi F, Abu Hassan M, Suleiman M: Feedback stabilization and adaptive stabilization of stochastic nonlinear systems by the control Lyapunov function. Stoch. Int. J. Probab. Stoch. Process. 2011, 83(2):179-201. 10.1080/17442508.2011.552723

2. Abedi F, Abu Hassan M, Arifin MN: Lyapunov function for nonuniform in time global asymptotic stability in probability with application to feedback stabilization. Acta Appl. Math. 2011, 116(1):107-117. 10.1007/s10440-011-9632-8

3. Abedi F, Leong JW: Dynamic robust stabilization of stochastic differential control systems. IMA J. Math. Control Inf. 2013, 30(4):559-569. 10.1093/imamci/dns040

4. Chen W, Jiao CL: Finite-time stability theorem of stochastic nonlinear systems. Automatica 2010, 46: 2105-2108. 10.1016/j. automatica. 2010.08.009

5. Deng H, Krstic M, Williams JR: Stabilization of stochastic nonlinear systems driven by noise of unknown covariance. IEEE Trans. Autom. Control 2001, 46(8):1237-1253. 10.1109/9.940927

6. Florchinger P: Lyapunov-like techniques for stochastic stability. SIAM J. Control Optim. 1995, 33: 1151-1169. 10.1137/S0363012993252309

7. Handel VR: Almost global stochastic stability. SIAM J. Control Optim. 2006, 45: 1297-1313. 10.1137/040618850

8. Khasminskii ZR: Stochastic Stability of Differential Equation. Sijthoff & Noordhoff, Alphen aan den Rijn; 1980.

9. Kushner JH: Converse theorems for stochastic Lyapunov functions. SIAM J. Control Optim. 1967, 5: 228-233. 10.1137/0305015

10. Lin Z, Liu J, Lin Y, Zhang W: Nonlinear stochastic passivity, feedback equivalence and global stabilization. Int. J. Robust Nonlinear Control 2012, 22: 999-1018. 10.1002/rnc.1742

11. Spiliotis J, Tsinias J: Notions of exponential robust stochastic stability, ISS and their Lyapunov characterizations. Int. J. Robust Nonlinear Control 2003, 13: 173-187. 10.1002/rnc.724

12. Trinh H, Fernando T: Functional Observers for Dynamical Systems. Springer, Berlin; 2012.

13. Xie XJ, Tian J: State-feedback stabilization for high-order stochastic nonlinear systems with stochastic inverse dynamics. Int. J. Robust Nonlinear Control 2007, 17(14):1343-1362. 10.1002/rnc.1177

14. Lan TTT, Dang HN: Exponential stability of nontrivial solutions of stochastic differential equations. Scientia, Ser. A, Math. Sci. 2011, 21: 97-106.

Chapter 3

Predicting the Probability of Failure of Cementitious Sewer Pipes Using Stochastic Finite Element Method

Amir M. Alani [1] **and Asaad Faramarzi** [2,*]

[1] School of Computing and Technology, University of West London, 8th Floor, Villiers House, Ealing Broadway, London W5 2PA, UK

[2] School of Civil Engineering, University of Birmingham, Edgbaston, Birmingham B15 2TT, UK

ABSTRACT

In this paper, a stochastic finite element method (SFEM) is employed to investigate the probability of failure of cementitious buried sewer pipes subjected to combined effect of corrosion and stresses. A non-linear time-dependant model is used to determine the extent of concrete corrosion. Using the SFEM, the effects of different random variables, including loads, pipe material, and corrosion on the remaining safe life of the cementitious sewer pipes are explored. A numerical example is presented to demonstrate the merit of the proposed SFEM in evaluating the effects of the contributing parameters upon the probability of failure of cementitious sewer pipes. The developed SFEM offers many advantages over traditional probabilistic techniques since it does not use any empirical equations in order to determine failure of pipes. The results of the SFEM can help the concerning industry (e.g., water companies) to better plan their resources by providing accurate prediction for the remaining safe life of cementitious sewer pipes.

KEYWORDS

stochastic finite element method; cementitious sewer pipe; probability of failure; concrete corrosion; random variables

1. INTRODUCTION

Underground sewer pipes are required to resist and operate safely under various external loads, as well as severe environmental conditions. The degradation of sewer pipes over time in combination with the effects of overlaying soil and surface traffic loads can sometimes cause catastrophic failures in sewer pipes. Despite the ever increasing spending on prevention and mitigation, there are thousands of collapse reports per year in sewer pipe networks throughout the world. The consequences of collapses of sewers are socially, economically, and environmentally devastating, causing, e.g., enormous disruption of daily life, massive costs, widespread pollution, and so on. It is known that sewer collapses are predominantly caused by deterioration of the pipes. For cementitious sewer pipes, corrosion is the main cause of deterioration. Corrosion can cause reduction in structural strength of the pipeline, leading to pipe collapse. Severe localised corrosion can cause pitting; also resulting in the failure of sewer pipes [1,2,3,4]. Therefore, considering the effect of corrosion in the analysis and design of cementitious sewer pipes is an essential element in developing an advanced model to predict the likelihood of collapses of sewer systems. Modelling corrosion in the structural analysis of a pipeline and simulating the performance of the pipe under various external load conditions together with the effect of corrosion provide a practical tool for both designer and asset managers to predict the service life or remaining safe life of the system.

A number of research studies in literature are devoted to the probabilistic analysis and reliability of underground pipelines. Ahammed and Melchers [5] derived a formula to estimate the maximum circumferential stress in underground, pressurised pipes subjected to external loading and corrosion. They combined a nonlinear time-dependant function for modelling the corrosion, and empirical equations proposed by Marston and Spangler [6,7] to estimate circumferential stresses in pipes due to the soil self-weight and traffic load. Then, using the obtained formula, they performed a reliability analysis to evaluate the probability of the failure of sewer pipes versus their service life. Ahammed and Melchers [8] extended their work to evaluate the reliability of underground, pressurised pipes, subjected to both circumferential and longitudinal stresses and corrosion. The equations developed by Marston and Spangler and used by Ahammed and Melchers [5,8] can provide estimations for stresses in the pipes due to external loads. However, these equations contain a number of empirical coefficients, which are usually estimated on the basis of limited information. The accuracy of the selected coefficients can significantly affect the accuracy of the predicted stresses in the pipe. Tanaka et al. [9] presented a mathematical model to find the survival probability of underground water pipelines. They demonstrated the application of the model to the water distribution network of Osaka, Japan. They also used the model to find the optimum renewal interval time for the different types of pipes in the network. Li et al. [10] used the Monte Carlo simulation technique to calculate the remaining safe life of metallic underground gas pipelines. They performed a sensitivity analysis to identify the key corrosion parameters in terms of pipeline failure. Lee et al. [11] presented the application of probabilistic modelling to evaluate the long-term performance of pipes. They evaluated the reliability of deteriorated pipes that were subjected to internal fluid pressure, external soil pressure and traffic loading. These studies and many similar works within literature have used empirical equations to estimate the stresses in the pressurised pipes. Alternative approaches, based on robust theoretical and numerical methods, are required to provide a more accurate estimation of the stresses in the pipes due to external loading and corrosion. In addition, there is a lack of research work in the area of reliability of cementitious sewers. It is acknowledged that H2S-induced (Hydrogen Sulphide) corrosion for cementitious pipes may have been well-researched but little is known about the combined effects of corrosion and external pressures on their remaining safe life in a quantitative manner. This may lead to a situation whereby some pipes may be replaced prematurely whilst others should be replaced but have not been. The former obviously is not cost-effective and the latter potentially presents undesirable challenge with devastating consequences. The developed model in this paper is an attempt to effectively eliminate this situation by accurately predicting pipes' remaining safe life, and ultimately contributing towards a more effective tool for planning.

In order to provide an accurate prediction of the remaining safe life of the sewer pipes, the parameters that affect and control the process of deterioration and failure of pipes, the interaction of different mechanisms of failure, and their effect on the remaining safe life of sewer pipes should be considered. Due to the large degree of uncertainty relating to the factors that are involved in the operation of underground sewer systems—in particular corrosion—it is probably more rational to model the failure of sewer pipes as a stochastic process. To fulfil this, a model has been developed and is reported in this paper that predicts the failure of cementitious sewer pipes using a stochastic finite element method. The developed stochastic finite element model can determine the prospect of failure of sewer pipes throughout their intended service life. Particular attention is paid to simulate the corrosion using the stochastic finite element model and to investigate its interaction with other mechanisms of failure and their effect on the remaining safe life of sewer pipes. Moreover, using the developed stochastic finite element model a parametric study is carried out to evaluate the significance and effect of each contributing parameter on the remaining safe life of the cementitious sewer pipes. The results provided by the developed stochastic finite element model can help asset managers and owners to make risk-informed and cost-minimised decisions with respect to when, where, what, and how interventions are required to ensure the safety and integrity of existing pipelines during their whole life of service.

2. STOCHASTIC FINITE ELEMENT METHOD

Stochastic finite element method (SFEM) is a powerful numerical tool in computational stochastic mechanics. SFEM can be classified as an extension of the classic deterministic finite element approach to the stochastic framework, i.e., to the solution of static and dynamic problems with stochastic mechanical, geometric, and/or loading properties [12].

The finite element method is a numerical technique used to obtain solutions to a wide range of partial differential equations describing various engineering problems including aerospace, geotechnical, structural engineering and many others. Most of the problems in engineering design and analysis can be modelled as a single or a set of differential equations. These differential equations describe the response of a system subjected to external influences. Many differential equations cannot be solved analytically and usually numerical techniques are used to find their approximate solutions. Of these numerical techniques the finite element method is known to be one of the most powerful general techniques for the numerical solution of a variety of problems encountered in engineering. The basic idea behind the finite element method is to divide the structure, body, or region being analysed into a large number of elements [13]. For a linear static analysis of a system the finite element method results in the following system of equations:

$$\mathbf{KU} = \mathbf{F} \tag{1}$$

where K is the global stiffness matrix and U and F represent the nodal displacement and force vectors. The global stiffness matrix (K) is obtained by adding the stiffness matrices of all elements,

$$\mathbf{K} = \sum \mathbf{K}^e \tag{2}$$

where Ke is:

$$\mathbf{K}^e = \int_{\Omega} \mathbf{B}^T \mathbf{D} \mathbf{B} \, d\Omega \tag{3}$$

In the above equation B is the strain matrix, D is the elasticity tensor, and Ω is elemental domain. The system of equations is solved to obtain the nodal displacements U and eventually other parameters such as stresses, strains etc. When one or some of the variables of the model (i.e., the loading and/or material properties and geometries) involve uncertainty, Equation (1) can be written in the following form [14,15]:

$$\mathbf{K}(\omega)\mathbf{U}(\omega) = \mathbf{F}(\omega) \tag{4}$$

where ω represents the randomness of the parameters. The above equation is a stochastic representation of the static finite element problems and the uncertain response of structure (i.e., U(ω)) and other quantities of interest such as stresses σ(ω), and strains ε(ω) can be obtained by solving Equation (4).

Although there are a number of variants of SFEM in the literature, it has been proven that Monte Carlo simulation (MCS) SFEM is by far the most general and powerful method to estimate the uncertainty of a structure's response [12,14,15,16,17,18,19]. In the MCS-based SFEM, a deterministic finite element problem is solved a large number of times and the response variability is calculated using statistical relationships. The MCS method does not involve any simplification or assumption, which makes it a robust and universal technique to treat complex SFEM problems. The only difficulty of MCS could be the significant computational cost that is required to perform a large number of simulations particularly when analysing problems with low probabilities of failure. However the recent advances and improvements in computer hardware and the use of efficient sampling methods to reduce the number of simulations, such as importance sampling [20], subset simulation [21], line sampling [22], local averaging [23,24], Latin-hypercube [25], and many more, have enabled the MCS-based SFEM to become a strong tool in the performance of the probabilistic analysis of various finite element models with uncertainty in load, displacement, and material properties.

MCS-based SFEM involves the generation of random numbers (random variables) with a uniform distribution between 0 and 1. Then the inverse cumulative probability function of the random numbers is used to produce the random numbers with arbitrary distribution function. The finite element model is created using both the deterministic parameters and the generated random variables. Limit state function(s) Z(X) of the problem is defined such that Z(X) ≤ 0 shows failure and positive values of Z(X) indicate no failure region. At every simulation, the limit state function(s) is checked using the finite element method (e.g., if the resultant stress has exceeded the yield stress) and the probability of failure is obtained using the following equation [20]:

$$P_f^i = \frac{N_f}{N} \tag{5}$$

where P i f is defined as the probability of failure of each limit state function, Nf is the number of simulations when Z(X) ≤ 0, and N is the total number of simulations. In a series system, where more than one limit state function exists, the failure of any of the limit state functions implies the failure of the system. If the individual failures are mutually independent, then the probability of the system can be obtained from:

$$P_f = 1 - \prod_{i=1}^{m} (1 - P_f^i) \tag{6}$$

where Pf is the probability of failure of the system, and is the number of limit state functions defined for the system.

In this paper an in-house SFEM code based on the integration of the finite element method and Monte Carlo simulation technique is developed to analyse the probability of failure of underground cementitious sewer pipes. The finite element code is equipped with a library of different 2D and 3D elements and a relatively fast solver to handle large size meshes. In the MCS code, the Latin-hypercube sampling technique is used to generate the random numbers with arbitrary inverse cumulative probability functions. In addition, the developed code is employed incrementally over the time in order to account for the degradation of the sewer pipe and predict the probability of failure of the sewer pipes throughout their service life.

In the SFEM, each uncertain parameter is represented by a random variable with a known probability density function, a mean value and a coefficient of variation. The random variables together with the deterministic parameters are used to generate the input file of the finite element model. The finite element analysis is performed and at the end of each simulation the failure criteria are checked. The failure criteria adopted in this study for the reliability analysis of cementitious sewer pipes are (i) the loss of strength of pipe; a Drucker-Prager yield criterion has been adopted to check if the yield strength of the pipe material has been reached, and (ii) excessive reduction of the pipe wall thickness due to the corrosion; the amount of corrosion is checked at every simulation step against a minimum allowable pipe wall thickness (this is explained in further detail in the next section). Each failure criterion is expressed as a limit state function and the probability of failure of the whole system is calculated using Equation (6).

3. EFFECT OF CORROSION

Concrete corrosion due to sulphuric acid attack is known to be one of the main contributory factors in the degradation of concrete sewer pipes. Sulphate, which exists in wastewater, is reduced to sulphide by anaerobic bacteria. These bacteria are present in a thin slime layer on the submerged surface of the sewer pipe and the production of sulphide occurs in this slime layer. The generated sulphide escapes to the exposed sewer atmosphere where it is transformed to sulphuric acid by aerobic bacteria. The acid reacts with the cementitious sewer pipe, which forms gypsum and causes corrosion [26,27,28]. Pomeroy [27] proposed an equation to predict the corrosion rate in sewer pipes. The proposed equation by Pomeroy [27] did not take into account the effect of the age of the pipe on corrosion rate. However a brief study of the available data on the corrosion of cementitious sewer pipes in literature can confirm that the loss of wall thickness does not occur at a constant rate over the design life of the pipe [29]. Therefore, in order to tackle this issue, in this study it is assumed that the corrosion of concrete sewer pipes is dependent on the age of the pipe and can be modelled in a practical engineering way to account for a time dependant corrosion process using a power law model. This approach has also been used by other research studies to model the corrosion of underground water and gas pipelines [5,8,30]. The power law model to predict the biogenic sulphuric acid corrosion of concrete pipes can be presented in the form of the following equation:

$$C = \alpha T^{\lambda} \tag{7}$$

where C is the corrosion of the pipe, α is a multiplying coefficient, λ is an exponential coefficient, and T is the age of the pipe. The data provided in Meyer [29] are utilised to evaluate the coefficients in Equation (7) using an exponential regression. Due to a large degree of uncertainty relating to the corrosion process, the corrosion and consequently the coefficients in Equation (7) are considered as stochastic parameters.

The predictions provided by Equation (7) are used in two parts of the simulation process. First, they are used to check whether the corrosion limit state function is violated or not. If the value provided by Equation (7) has exceeded the minimum allowable wall thickness then it is recorded as a corrosion failure and will be used to ascertain the probability of failure due to corrosion (Equation (5)). Equation (7) is also used to update the finite element mesh of the pipe at every simulation to represent the reduction of the wall thickness due to corrosion. It is worth mentioning that corrosion only attacks the exposed surface of the sewer pipe, above the sewage, with its deepest penetration in the crown of the pipe [27]. Although the actual distribution of the corrosion in a concrete sewer pipe is unequal (Figure 1a), in this study it is assumed that the corrosion has a uniform distribution (Figure 1b). For every finite element simulation the amount of corrosion is obtained using Equation (7) and the finite element input file is updated by re-meshing the pipe domain using the new coordinates after considering the corrosion. The strength of the concrete pipe is expected to be reduced as a result of the reduction of pipe wall thickness.

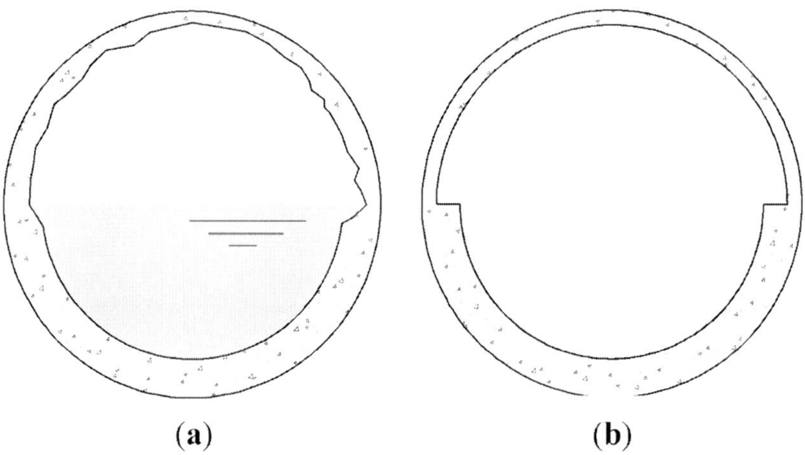

Figure 1. Corrosion in concrete sewer pipes (a) Realistic distribution of corrosion in a concrete sewer pipe (non-uniform); (b) Idealised corrosion of concrete sewer pipes in the SFEM (uniform).

4. NUMERICAL EXAMPLE

In this section a numerical example is considered to evaluate the performance of the developed SFEM in predicting the probability of failure of concrete sewer pipes subjected to stresses and corrosion. The presented example is also used to investigate the influence of the contributing parameters in the remaining safe life of the sewer pipes and their interaction with each other. The finite element (FE) model of the problem consists of a concrete pipe with a circular section buried underground and surrounded by a homogenous soil. The model is subjected to self-weight and an external traffic load applied on the surface of the model. The FE model is assumed to be two dimensional with plane strain geometrical condition (i.e., 2D plane strain elements are used to create the FE mesh). In order to draw conclusions that are not affected by a particular example, the problem is scaled with respect to the external diameter of the pipe and the variations of different normalised parameters are investigated. Figure 2 shows the geometry, boundary conditions, and a typical FE mesh of the problem.

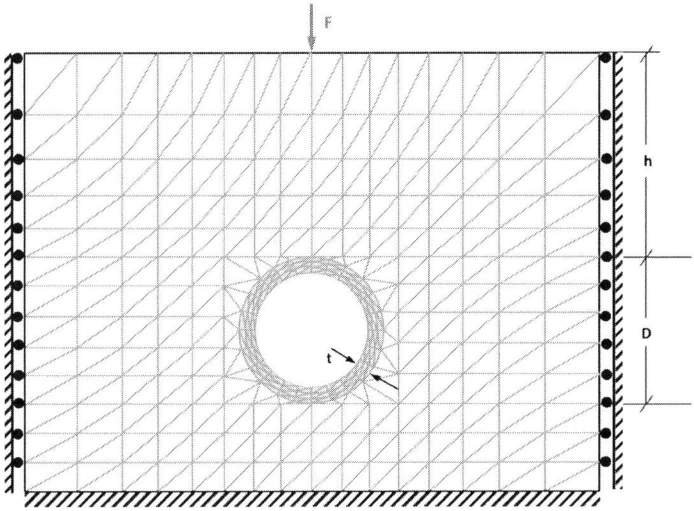

Figure 2. Geometry, mesh and parameters used in the FE model.

The normalised parameters are expressed in Equations (8–10).

$$h^* = \frac{h}{D} \tag{8}$$

$$t^* = \frac{t}{D} \tag{9}$$

$$F^* = \frac{F}{\gamma h D^2} \tag{10}$$

In the above equations, h is the height of the soil under which the pipe is buried, t is the initial thickness of pipe, D is the external diameter of the pipe, F is the traffic wheel load, and γ is the unit weight of the soil.

In addition to the deterministic parameters such as diameter and thickness of the pipe, there are some parameters that are considered as stochastic or random. Using the existing studies on the reliability analyses of underground pipelines (e.g., [5,8]) and, after performing a number of pilot simulations, five parameters were chosen as random variables. These parameters, their mean and their coefficient of variation (cv) are presented in Table 1. The time-dependant growth of the traffic load over the lifetime of the pipe has been ignored in this study. Other parameters, such as unit weight of soils, were considered deterministic mainly because their coefficients of variation were very insignificant or they had small or negligible effects on the probability of failure of the system.

Table 1. Random variables used for the SFEM.

Symbol	Description	Mean	Coefficient of Variation
f_c	Concrete maximum compressive strength	35 MPa	0.1
F^*	Traffic Load	Varies from 10–20	0.25
I_f	Load impact factor	1.5	0.15
α	Corrosion multiplying coefficient	3.5×10^{-5}	0.1
λ	Corrosion exponential coefficient	1.5	0.15

5. RESULTS AND DISCUSSION

After choosing the type of each parameter (deterministic or stochastic) and creating the FE models, the time-dependant SFEM were performed for different values of scaled parameters in order to study their effect on the probability of failure of concrete sewer pipes. Figure 3a shows the results of the SFEM on probability of failure of the example sewer pipe under various normalised traffic loads over its service life. It can be seen that, as the traffic load increases, the probability of failure of sewer pipes grows rapidly. In addition it can be noted that initially the probability of failure is zero or very small for all cases; however as the effect of corrosion emerges (usually after the first 20 years) the probability of failure increases rapidly. To further investigate the effect of traffic load on the service life of concrete sewer pipes subjected to stresses and corrosion, the following analysis was also carried out. Let us assume that the acceptable probability of failure (Pf) is 10% (equivalent of a remaining safe life of 90%). The service life of each FE model (each model has a different normalised traffic load) can be evaluated using the results provided by the SFEM (Figure 3b). It can be seen in Figure 3b that the service life of the sewer pipe is reduced significantly with a non-linear trend as the traffic load increases. For example if the traffic load is doubled (i.e., F* increases from 10 to 20) then the service life of the sewer pipe is reduced from 60 years to zero years for the acceptable probability of failure of 10%. A similar trend can also be seen for other presumed acceptable values of probability of failure (i.e., Pf = 20% and Pf = 30%).

The effect of the buried depth of the sewer pipe on the probability of failure is shown in Figure 4a. The figure shows that reduction in soil cover can result in an increase in the probability of failure of sewer pipes with a nonlinear relationship. In Figure 4b, it is shown that an increase of 100% of soil cover height can extend the service life of the sewer pipe by at least an extra 30 years. The reduction in probability of failure of sewer pipes for higher values of h* can be attributed to the fact that at greater depths, the effect of traffic load becomes smaller compared to the corrosion, particularly as the age of the pipe increases. This results in a situation where the corrosion failure may become the main mechanism of failure. As can be noted from Equation (7), corrosion is not a function of depth and the same values are applicable for both deep and shallow pipes. Further details can be seen in Figure 5. In Figure 5 the probability of failure due to corrosion, stress and total failure are presented for two different depths. It can be noted that the main cause of the failure of sewer pipe for shallow depth, h* = 1.5 (Figure 5a), is the stress throughout the service life of the pipe while for h* = 3.0 (Figure 5b) corrosion becomes the dominant mechanism of failure after 90 years of service life. Similar conclusions to those obtained for the effect of traffic load and cover soil on probability of failure can be drawn for the thickness of the pipe (see Figure 6a,b).

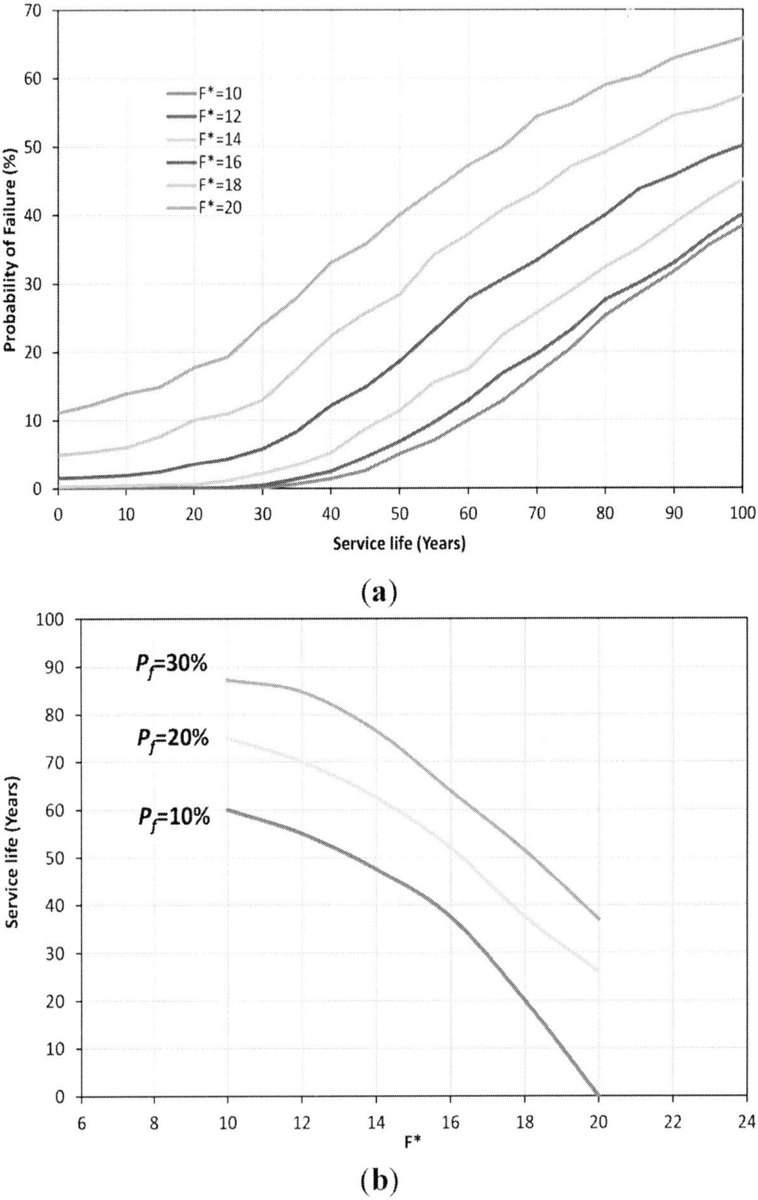

Figure 3. (a) Probability of failure of concrete sewer pipe for different values of scaled traffic load versus service-life; (b) The variation of traffic load (normalised) with the service life of sewer pipes.

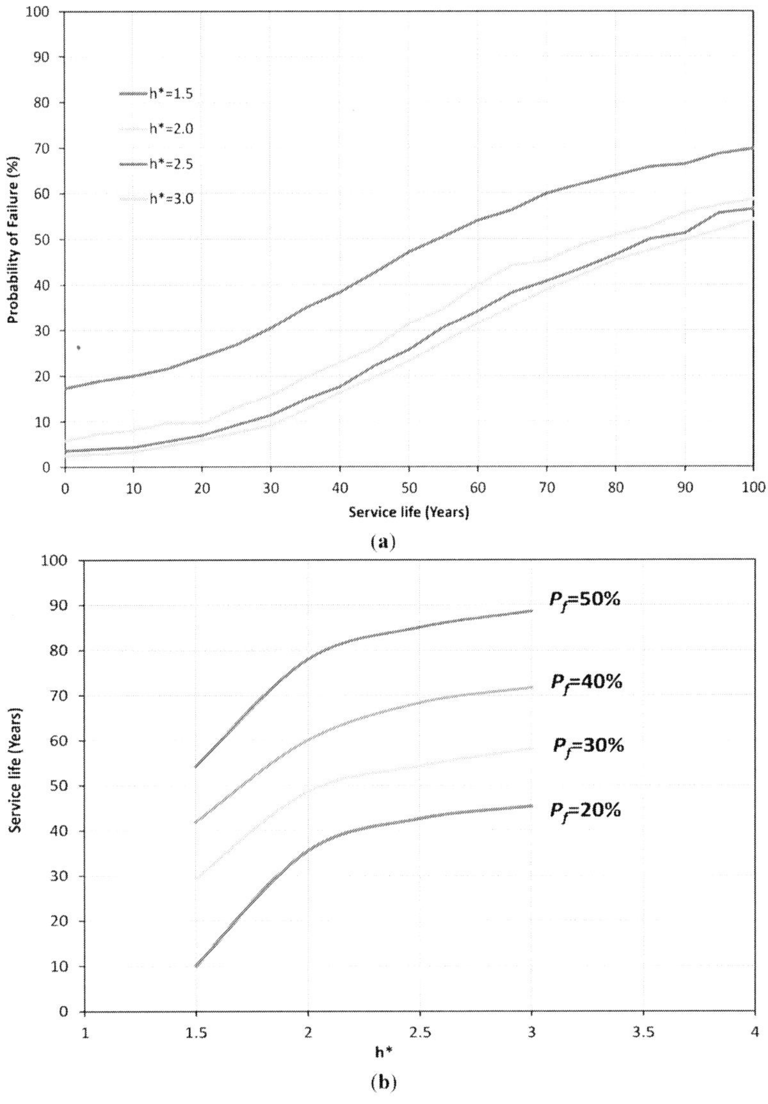

Figure 4. (a) Probability of failure of concrete sewer pipe for different values of h*; (b) The variation of service life of sewer pipe versus different h*.

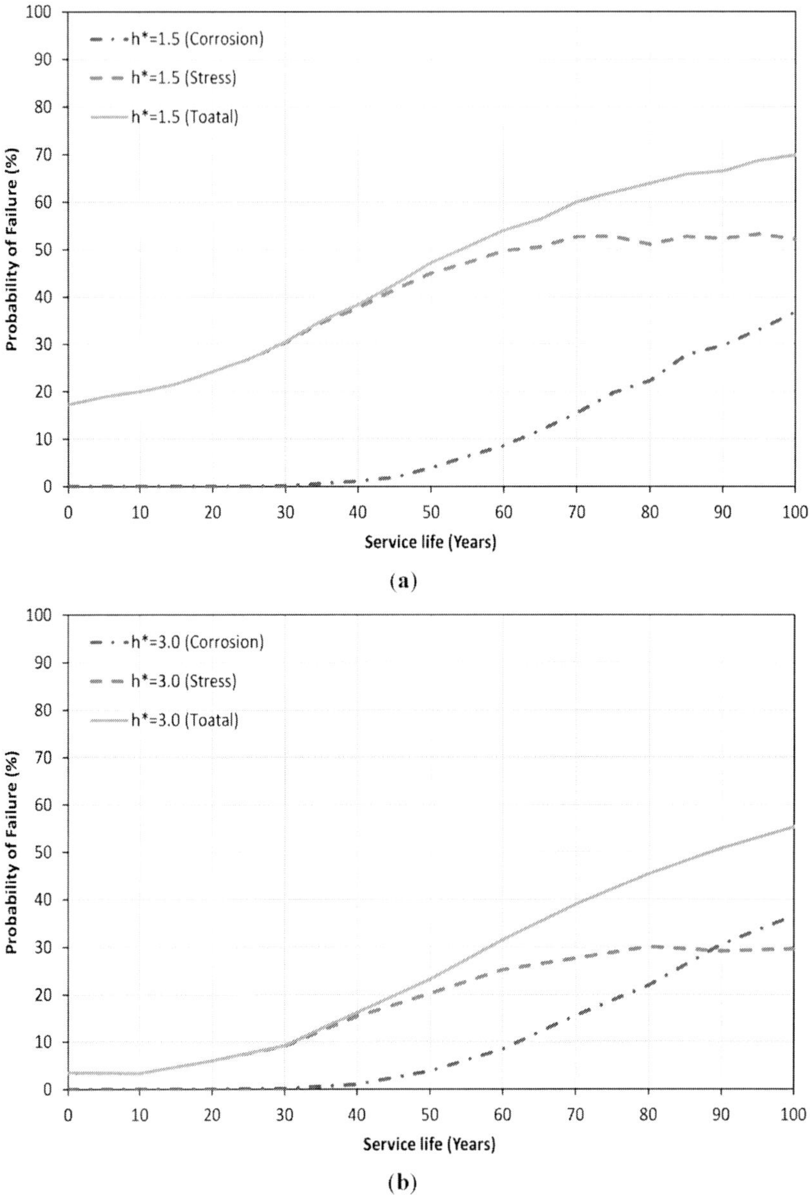

Figure 5. Probability of failure due to corrosion, stress and total failure for (a) h* = 1.5; (b) h* = 3.0.

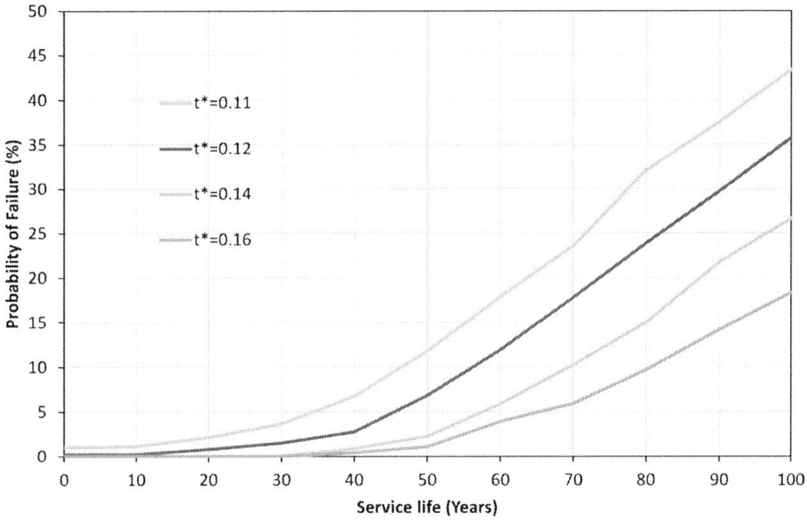

(a)

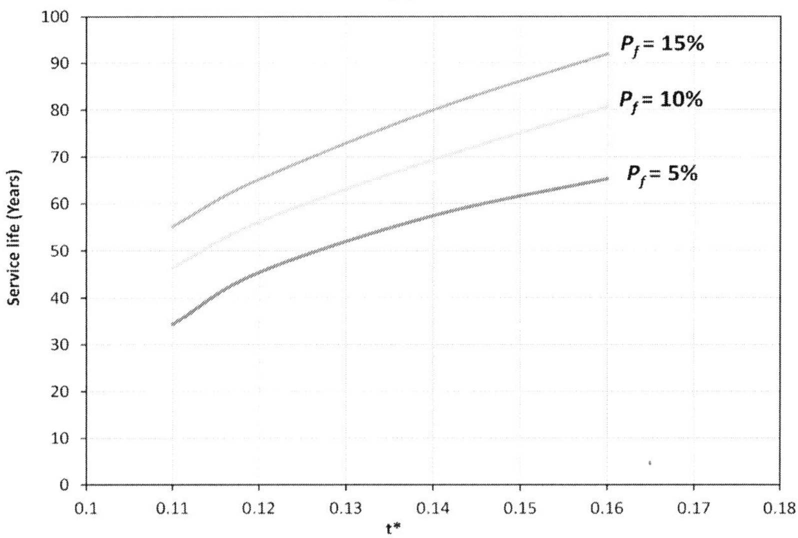

(b)

Figure 6. (a) Probability of failure of concrete sewer pipe versus its service life for different values of normalised pipe thickness; (b) The variation of service life of sewer pipe versus different t*.

Figure 3, Figure 4, Figure 5 and Figure 6 show the effect of both deterministic (i.e., h* and t*) and stochastic (F*) parameters on the probability of the failure of sewer pipes. In order to investigate the relative contribution and importance of each stochastic (random) variable in the probability of failure of concrete sewer pipes, a sensitivity analysis was carried out. The results of the sensitivity analyses for two examples with normalised soil cover (h*) after 100 years of service life are presented in Figure 7a,b. In both figures it can be seen that the exponential coefficient of corrosion extensively contributes to the probability of the failure of concrete sewer pipes. It can also be noted that this contribution enlarges as the soil cover increases. As previously explained, this is mainly due to the fact that, at higher values of soil cover, the dominant failure mode can become the corrosion failure as the pipe passes a certain age. A further point to discuss is the importance of the models that predict the corrosion in sewer pipes. For example, at a scaled buried depth of h* = 3.0 it can be seen that the corrosion coefficients (i.e., α, λ) together contribute to nearly half (48%) of the probability of failure of sewer pipes. This shows the significant influence of corrosion in general and corrosion coefficients in particular on the prediction of probability of failure. In general, the results of the sensitivity analysis can be used to efficiently improve the design towards a higher reliability for sewer pipe projects or to enhance management, rehabilitation and spending on existing pipelines.

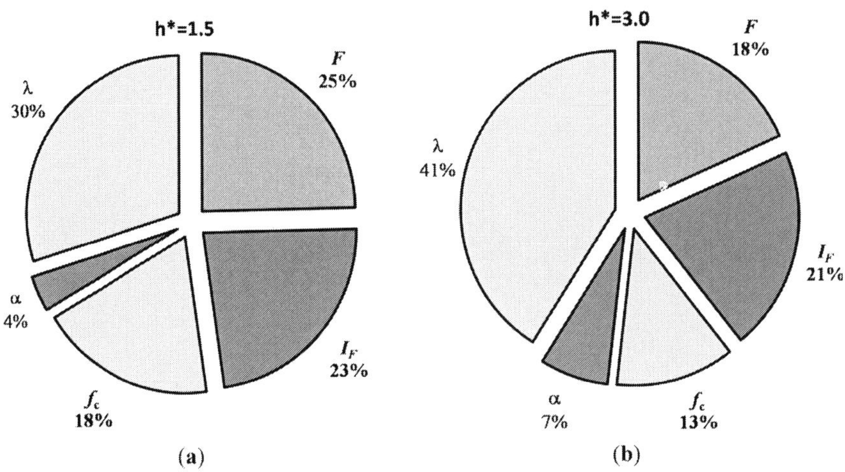

Figure 7. Relative contribution of random variables to the probability of failure of sewer pipe after 100 years of service life (a) h* = 1.5; (b) h* = 3.0.

6. SUMMARY AND CONCLUSIONS

In the present study a stochastic finite element method was utilised to predict the probability of failure of concrete sewer pipes subjected to combined effects of corrosion and stresses. Uncertainties involved in pipe material, traffic load, and corrosion are considered to develop the stochastic finite element model. A nonlinear time-dependent model was chosen to predict the corrosion in concrete sewer pipes. The model parameters were chosen based on a set of existing data in the literature. A normalised numerical example was employed to investigate the effect of both deterministic and probabilistic parameters on the probability of failure of sewer pipes. Two mechanisms of failure (i.e., corrosion and stress failure) were adopted to define the limit state functions. The results of the numerical simulations revealed a nonlinear relationship between most of the parameters and the probability of failure of sewer pipes. In addition the results of the sensitivity analyses showed the significant contribution of the corrosion parameters. Study of the available literature clearly demonstrated that there is a serious lack of monitoring data of underground sewer pipes. The monitoring field data compiled by water industry are neither readily available nor comprehensive enough to be utilised for appropriate analytical purposes. No doubt access to such data is of paramount importance to check the validity of developed model in this study. The authors are currently working with concerning industries in order to have access or compile the much-needed data.

The results of the SFEM can be used to improve the performance and planning of existing sewer systems, by providing better predictions for the probability of failure of sewer pipes compared to the existing approaches. The SFEM can bring together the effects of contributing parameters in the probability of failure of the system being studied in a numerical framework with high precision. Using the SFEM, it is possible to study the effect of each parameter on the failure of the system and their interaction with each other. The SFEM also provides a time-dependant reliability analysis for predicting the remaining safe life of sewers, which provides a means to better manage the existing sewers and plan resources during their whole life of service.

ACKNOWLEDGMENTS

This research was funded by a grant from the UK Engineering and Physical Sciences Research Council (EPSRC), grant number EP/1032150/1 (Assessing Current State of Buried Sewer Systems and Their Remaining Safe Life).

AUTHOR CONTRIBUTIONS

Asaad Faramarzi developed the numerical model and wrote the manuscript. Amir M. Alani contributed towards the development and validation of the model and edited the manuscript. These authors contributed equally to this work.

REFERENCES

1. Angerville, R.; Perrodin, Y.; Bazin, C.; Emmanuel, E. Evaluation of Ecotoxicological Risks Related to the Discharge of Combined Sewer Overflows (CSOs) in a Periurban River. Int. J. Environ. Res. Public Health 2013, 10, 2670–2687.

2. Fobil, J.; May, J.; Kraemer, A.; Emmanuel, E. Assessing the Relationship between Socioeconomic Conditions and Urban Environmental Quality in Accra, Ghana. Int. J. Environ. Res. Public Health 2010, 7, 125–145.

3. Alani, A.M.; Faramarzi, A. An evolutionary approach to modelling concrete degradation due to sulphuric acid attack. Appl. Soft Comput. 2014, 24, 985–993.

4. Alani, A.M.; Faramarzi, A.; Mahmoodian, M.; Tee, K.F. Prediction of sulphide build-up in filled sewer pipes. Environ. Technol. 2014, 35, 1721–1728.

5. Ahammed, M.; Melchers, R.E. Reliability of underground pipelines subject to corrosion. J. Transp. Eng. 1994, 120, 989–1002.

6. Spangler, M.G.; Handy, R.L. Soil Engineering, 4th ed.; Harper and Row: New York, NY, USA, 1982.

7. Moser, A.P.; Folkman, S. Buried Pipe Design, 3rd ed.; McGraw-Hill: New York, NY, USA, 1998.

8. Ahammed, M.; Melchers, R.E. Probabilistic analysis of underground pipelines subject to combined stresses and corrosion. Eng. Struct. 1998, 19, 988–994.

9. Tanaka, T.; Nam, L.T.; Kaito, K.; Kobayashi, K. Probabilistic analysis of underground pipelines for optimal renewal time. J. Water Supply Res. Technol.-AQUA 2010, 59, 445–451.

10. Li, S.; Yu, S.; Zeng, H.; Li, J.; Liang, R. Predicting corrosion remaining life of underground pipelines with a mechanically-based probabilistic model. J. Pet. Sci. Eng. 2009, 65, 162–166.

11. Lee, L.S.; Estrada, H.; Baumert, M. Time-Dependent Reliability Analysis of FRP Rehabilitated Pipes. J. Compos. Constr. 2010, 14, 272–279.

12. Stefanou, G. The stochastic finite element method: Past, present and future. Comput. Methods Appl. Mech. Eng. 2009, 198, 1031–1051.

13. Stasa, F. Applied Finite Element Analysis for Engineers; Holt, Rinehart & Winston of Canada: Toronto, ON, Canada, 1986.

14. Pellissetti, M.F.; Schuëller, G.I. Scalable uncertainty and reliability analysis by integration of advanced Monte Carlo simulation and generic finite element solvers. Comput. Struct. 2009, 87, 930–947.

15. Kamiński, M. The Stochastic Perturbation Method for Computational Mechanics, 1st ed.; John Wiley & Sons: Chichester, UK, 2013.

16. Yamazaki, F.; Shinozuka, M.; Dasgupta, G. Neumann expansion for stochastic finite element analysis. J. Eng. Mech. 1988, 114, 1335–1354.

17. Papadrakakis, M.; Papadopoulos, V. Robust and efficient methods for stochastic finite element analysis using Monte Carlo simulation. Comput. Methods Appl. Mech. Eng. 1996, 134, 325–340.

18. Shang, S.; Yun, G. Stochastic finite element with material uncertainties: Implementation in a general purpose simulation program. Finite Elem. Anal. Des. 2013, 64, 65–78.

19. Reh, S.; Beley, J.D.; Mukherjee, S.; Khor, E.H. Probabilistic finite element analysis using ANSYS. Struct. Saf. 2006, 28, 17–43. [Google Scholar] [CrossRef]

20. Melchers, R.E. Structural Reliability Analysis and Prediction, 2nd ed.; John Wiley & Sons: Chichester, UK, 1999.

21. Au, S.K.; Beck, J.L. Estimation of small failure probabilities in high dimensions by subset simulation. Probab. Eng. Mech. 2001, 16, 263–277.

22. Koutsourelakis, P.S.; Pradlwarter, H.J.; Schuëller, G.I. Reliability of structures in high dimensions, Part I: Algorithms and applications. Probab. Eng. Mech. 2004, 19, 409–417.

23. Griffiths, D.V.; Fenton, G. Probabilistic Slope Stability Analysis by Finite Elements. J. Geotech. Geoenviron. Eng. 2004, 130, 507–518.

24. Fenton, G.; Vanmarcke, E.H. Simulation of random fields via local average subdivision. J. Eng. Mech. 1990, 116, 1733–1749.

25. Florian, A. An efficient sampling scheme: Updated Latin Hypercube Sampling. Probab. Eng. Mech. 1992, 7, 123–130.

26. Pomeroy, R.D.; Parkhurst, J.D. The forecasting of sulfide build-up rates in sewers. Prog. Water Technol. 1977, 9, 621–628.

27. Pomeroy, R.D. The Problem of Hydrogen Sulphide in Sewers; Clay Pipe Development Association: London, UK, 1976.

28. Beeldens, A.; Van Gemert, D. Biogenic sulphuric acid attack of concrete sewer pipes: A prediction of the corrosion rate. In Proceedings of the 5th CANMET/ACI International Conference on Recent Advances in Concrete Technology, Singapore, 29 July–1 August 2001.

29. Meyer, W.J. Case study of prediction of sulphide generation and corrosion in sewers. Water Pollut. Control Fed. 1980, 52, 2666–2674. [Google Scholar]

30. Amirat, A.; Mohamed-Chateauneuf, A.; Chaoui, K. Reliability assessment of underground pipeline under the combined effect of active corrosion and residual stress. Int. J. Press. Vessel. Pip. 2006, 83, 107–117.

Chapter 4

Stochastic Process Optimization Technique

Hiroaki Yoshida[1], Katsuhito Yamaguchi[2], Yoshio Ishikawa[1]

[1]College of Science and Technology, Nihon University, Chiba, Japan

[2]Junior College, Nihon University, Chiba, Japan

ABSTRACT

The conventional optimization methods were generally based on a deterministic approach, since their purpose is to find out an accurate solution. However, when the solution space is extremely narrowed as a result of setting many inequality constraints, an ingenious scheme based on experience may be needed. Similarly, parameters must be adjusted with solution search algorithms when nonlinearity of the problem is strong, because the risk of falling into local solution is high. Thus, we here propose a new method in which the optimization problem is replaced with stochastic process based on path integral techniques used in quantum mechanics and an approximate value of optimal solution is calculated as an expected value instead of accurate value. It was checked through some optimization problems that this method using stochastic process is effective. We call this new optimization method "stochastic process optimization technique (SPOT)". It is expected that this method will enable efficient optimization by avoiding the above difficulties. In this report, a new optimization method based on a stochastic process is formulated, and several calculation examples are shown to prove its effectiveness as a method to obtain approximate solution for optimization problems.

KEYWORDS

Optimization, Stochastic Process, Path Integral, Expected Value, Quantum Mechanics

1. INTRODUCTION

Optimization methods are useful to system design, and especially the applications to engineering problems are expected. In order to obtain the accurate optimal solution, generally the conventional optimization methods were based on a deterministic approach. However, when design variables have many inequality restriction conditions, in order to search for a solution efficiently, it is necessary to adjust a parameter with trial and error. Moreover, if the problem is complex, the risk of falling into a local solution is also high.

Our method obtains an approximate value of an optimal solution using stochastic process. The optimization methods using stochastic process like simulated annealing (SA) [1] are found. However, in these methods, stochastic process is used as a means of searching for an accurate solution efficiently. Our method is based on probability theory, and essentially differs from the concepts of the conventional optimization methods. The purpose of the conventional optimization is to obtain the accurate optimal solution, but that of our method is to obtain the approximate value of the optimal solution.

We showed that the approximate value of the optimal solution could be obtained using stochastic process [2] [3] . And even if it was an approximate value of the optimal solution obtained by this method, it was shown that it is useful in engineering [4] . Furthermore, it was shown that this method can apply to general engineering optimization problems containing static design variables and dynamic design variables [5] .

As mentioned above, it was shown that the approximate value of the optimal solution can be obtained by this method using stochastic process. The aim of this paper is not to show the results of the applications of our method, but to bring our original idea to a wide reading audience.

Thus, we propose a new method in which an optimization problem is replaced with a stochastic process based on path integral techniques used in quantum mechanics and an approximate value of an optimal solution is calculated as an expected value instead of an accurate value. We call this new optimization method the "stochastic process optimization technique (SPOT)". It is expected that this method will enable efficient optimization by avoiding the above difficulties.

Since expected values are stochastic average values, a stochastic approach is not appropriate to obtain an accurate solution as a deterministic approach.

However, its characteristics are simple algorithms and the fewer number of manual parameters. The resultant solution is approximate enough to a global solution of engineering problems whose solution spaces are expected to be relatively smooth. The approximate solution can be an initial value to obtain an accurate solution with a deterministic process.

In this paper, a new optimization method is formulated based on a stochastic process. Some calculation examples are introduced to show how effective the approximate solutions are for optimization problems. The examples are a simple benchmark test, a classic line of swiftest descent problem, and an optimized glide problem of a hang glider as an introductory engineering optimization problem.

2. PATH INTEGRAL AND OPTIMIZATION

According to the principle of least action, Newtonian motion of a mass point makes the action integral minimum:

$$I = \int_{t_1}^{t_2} L\left(x_i, \mathrm{d}x_i/\mathrm{d}t, t\right)\mathrm{d}t \quad , (1)$$

where $L\left(x_i, \mathrm{d}x_i/\mathrm{d}t, t\right)$ is the Lagrange function. Also, the Euler-Lagrange equation obtained with variation of I is equal to an equation of Newtonian motion. On the other hand, the motion of a particle is determined stochastically according to the quantum mechanics. A motion on the classical mechanics, which keeps I minimum, occurs at the “highest possibility”. However, other motions can occur at certain probabilities. This results in some particular phenomena, such as the tunnel effect, which never occur on the classical mechanics.

The well-known simulated annealing (SA) method adopts such a quantum mechanical approach to the solution of optimization problems. SA introduces a probability distribution that reaches its maximum at an optimal value. This method numerically searches the optimal value (peak of the probability distribution) with a temperature parameter using the probability distribution that reaches its maximum of an optimal value instead of directly obtaining a variational problem of a performance index I. This method searches solution spaces in the large, which include solutions that never obtained with the variational method. Therefore, one advantage is a low probability of reaching local solutions. On the other hand, there are some disadvantages. For example, the operation of a cooling schedule, which is a temperature parameter, is difficult. In addition, it may take much time in calculation to converge a solution.

R. Feynman constructed the "path integral" theory in 1948 and showed that it is equivalent to the quantum mechanics. The theory provides a countless number of paths that meet given boundary conditions, and the prob- ability (probability amplitude) of each particle taking its path is expressed in the form of multiple integral [6] . A stochastic behavior is the nature of quantum mechanics. It is therefore impossible to predict a measured physical quantity deterministically under a certain condition. Expected values only can be predicted. Thus, the expected value of a physical quantity is expressed in the form of multiple integral in the path integral method. Furthermore, the multiple integral calculation of an expected value can fit into numerical calculation with the Monte Carlo method.

In this paper, we propose a new numerical calculation method for optimization problems based on the quantum mechanics rather than SA. It means that we show a method of obtaining an approximate solution of an optimal solution as an expected value. An expected value is predicted to be located close to an optimal solution in a continuous probability distribution. This method can thus be applied to various engineering optimal problems.

The introduced probability distribution P here is basically the same as that in SA, and to obtain the minimum of performance function I, the following equation is set:

$$P = e^{-I/h} \quad . (2)$$

In SA, the peak of P is searched with $h \to 0$. On the other hand, we automatically generate a state variable of various time histories at probability P as finite value h. According to the generated frequency, the stochastic average (expected value) of the quantity to be obtained is calculated.

3. FORMULATION OF SPOT

According to the path integral method, the path of motion for a naturally stochastic particle such as photon and electron is obtained from an expected value as a path that keeps the action integral I minimum. We replace the action integral I with a performance index to recognize the motion of a general object as a stochastic process. Thus, we formulate the path integral method for a naturally stochastic particle in physics as an optimal method of obtaining a function that keeps the performance function I minimum. Note that since this performance function is stochastic, an optimal solution is obtained as an expected value, which results in an approximate solution instead of an accurate solution.

According to the path integral method, the action integral $I[x(t)]$ described in the above section is replaced with a performance function to define probability distribution P as shown in Equation (3). Here, $x(t)$ does not always show a path, but shows a general variable to be optimized. Also, A is the normalization constant.

$$P = \frac{1}{A}\exp(-I/h) \quad , (3)$$

where h is the Plank constant in quantum mechanics. A stochastic mechanics system matches a classical mechanics system as $h \to 0$. However, in this situation, h is an arbitrary parameter that gives a fluctuation of solution. It means that h is a parameter to define a distribution of P. The larger h is, the wider the distribution is. Also, A is the normalization constant to make the summation of the probability 1. To define the probability distribution as shown in Equation (3), the expected value is obtained from the following:

$$\langle x \rangle = \int_{-\infty}^{\infty} \left[x(t) \cdot P \right] Dx(t) \quad , (4)$$

where $Dx(t) = \mathrm{d}x_0 \mathrm{d}x_1 \cdots \mathrm{d}x_N$ shows the summation for the paths.

Here, the summation for the paths is the integral for all the countless paths (multiple integral), which for example connect each point x_i with lines at each time point $t_i\ (0 \le i \le N)$ divided by N in one-dimensional motion. Figure 1 shows some paths that are generated in a probability distribution of Equation (3) when a function defined in Equation (1) is a performance index in one-dimensional upcast motion.

The horizontal axis and vertical axis show time and vertical positions, respectively. The better performance values the paths have, in other words, the closer to the accurate solution (red line) the values are, the higher the density of the paths is. Also, there are some paths that never exist with a deterministic method. The optimal solution is the expected value of all the paths. In this case, that solution is an approximate solution of an accurate solution. The expected solution that is thus obtained shows a path of the highest probability for move of a stochastic behavioral particle, such as an electron, between two points. When those paths are replaced with arbitrary functions, approximate solutions that make the performance function values minimum are obtained.

This idea can be extended to obtain an expected value of a m-dimensional functional that consists of $m \times N$ functions of time in the following way. First, independent variable t is divided by N, and $x_k^{(i)}$ values at time $t_i \ (0 \leq i \leq N)$ are randomly chosen. Next, a path that connects these points is set to the time history of variable x_k (Figure 2).

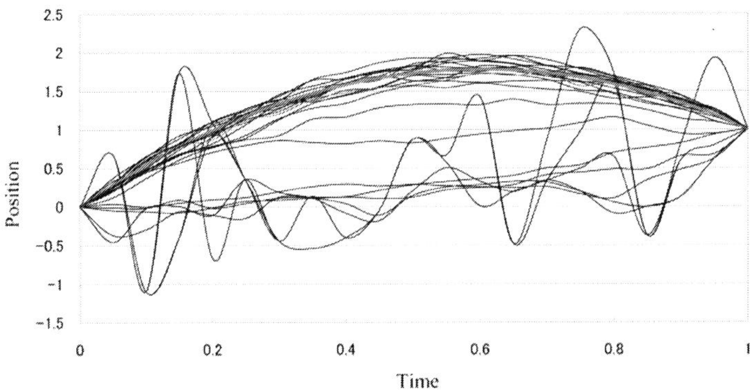

Figure 1. Paths generated by a probability distribution.

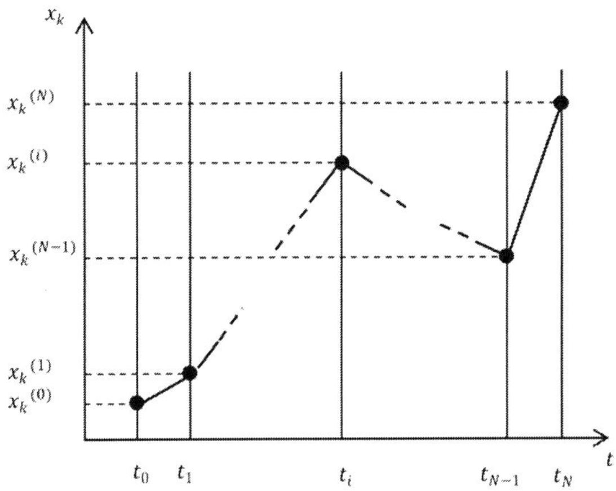

Figure 2. Time history of a variable.

Note that the subscript k shows the k-th variable $(1 \le k \le m)$. From Equation (4), the expected value $\langle x_k^{(i)} \rangle$ of x_k at t_i $(0 \le i \le N)$ is expressed as follows:

$$x_k^{(i)} = \int_{-\infty}^{\infty} \left[x_k^{(i)} \cdot P \right] Dx^{(0)}(t) Dx^{(1)}(t) \cdots Dx^{(N)}(t),$$
$$Dx^{(i)} = \mathrm{d}x_1^{(i)} \cdot \mathrm{d}x_2^{(i)} \cdots \mathrm{d}x_m^{(i)}. \quad (5)$$

The integral range is from $-\infty$ to ∞ in Equation (5) according to the form of the path integral. The range in actual calculation is between the upper and lower limits of each variable. Also, Equation (5) is formulated to set the values according to the probability distribution for variable $x_k^{(i)}$ at time 0 and N. However, a variable may be fixed in some problems. The end points should be set in accordance with the boundary conditions.

Note that since Equation (5) shows the multiple integral, the Monte Carlo method can be applied to actual numerical calculations.

4. CALCULATION ALGORITHM FOR SPOT

First, a performance index is determined, and if necessary, variables are discretized for numerical calculation. Next, the values of all variables are randomly chosen to generate solutions. Generation probabilities of variables should agree with the probability distribution of Equation (3). Thus, a solution with a fine performance value is generated at a high frequency, and a solution with a poor performance index value is generated at a low frequency. Also, the algorithm is based on the Markov process so that future solutions will have no dependency on the former solutions. Lastly, expected values are calculated using all the solutions generated with a probability distribution to obtain an approximate solution.

The procedure of calculation is shown in Figure 3.

1) The values of all the variables are generated randomly to obtain initial solutions.

2) Time point t_i is chosen according to a random number, and function value x_i at t_i is randomly rewritten to generate a different solution.

3) The performance index value of the newly generated solution is calculated.

4) The adoption or rejection of this solution is decided to generate solutions depending on the probability distribution of Equation (3).

The procedure from Steps 2) to 4) is a pure explanation of the Metropolis method [7] to merely conduct the multiple integral. This method has an advantage of no necessity to obtain the normalization constant: A in Equation (3). A prime characteristic of this method is to calculate the expected values in Step 5).

5) The expected value is calculated.

6) If the final condition is met, the calculation ends. If not, it returns to the generation of solutions.

The final condition can be a certain range of variation of an expected value. However, in this paper, the final condition is the number of iterations, $K_{\max}$. This is because the generation of solutions during the calculation is a Markov process, and therefore, the still more iterations of calculation can result in fine performance index values if the convergence of expected values is not enough.

Equation (3) for probability distribution in this paper is the same as that in SA. SA is a deterministic approach to search only optimal solution at $h \to 0$. On the other hand, h is finite and an optimal solution is obtained as an expected value (stochastic average value) in this paper. It means that the two approaches are based on fundamentally different concepts. In the concrete, SA starts a search with an initial solution for candidate solutions according to the probability distribution. It adopts a solution of the best performance function value as an optimal solution. On the other hand, our method determines an approximate solution to an optimal solution as an expected value with all the solutions generated according to the probability distribution. In this method, there is no process of the so-called cooling schedule which is in SA.

5. CALCULATION EXAMPLE

This section describes examples of numerical calculation with SPOT to demonstrate the validity. First, SPOT is applied to a problem where the minimum value of a multi-peaked function is obtained without staying at other minimal values and the minimum value is always only one with no dependency on the initial value. Next, we apply SPOT to a problem of obtaining the minimum value of a functional to show the validity in a physical optimization problem. Lastly, we apply SPOT to an engineering optimal problem on the glide of a hang glider to show the high applicability of an approximate solution by SPOT.

We already applied SPOT to some engineering problems [8] [9] .

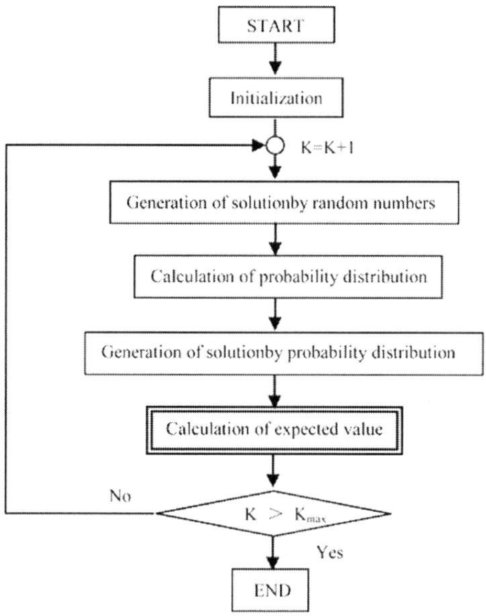

Figure 3. Flow chart of calculation.

5.1. Minimization Problem of Bivariate Function

First, we apply SPOT to a problem of obtaining the minimum value of a bivariate function expressed in Equation (6). Here, performance function I is the function value $f(x,y)$ as follows:

$$f(x,y)=\{\cos(2\pi x)+\cos(2.5\pi x)-2.1\}\times\{2.1-\cos(3\pi y)-\cos(3.5\pi y)\},\quad 0.0\le x\le 3.0,\ 0.0\le y\le 1.5 \tag{6}$$

Figure 4(a) shows the solution space of Equation (6). The set of $f(x,y)$ that minimizes this function is (0.439, 0.306). This function has multiple minimal values and thus is a multi-peaked function. The contours are shown in the bottom plane, and Figure 4(b) shows the top view.

This example is a problem for obtaining the minimum value of a bivariate function. Therefore, the variables are not discretized, but the values of two variables are generated according to the probability distribution (3) defined by the performance values to obtain an expected value. The closed dots in Figure 4(a) show multiple initial values chosen to check the dependency on the initial values. Also, the cross marks in Figure 4(b) show the minimum values obtained with this method.

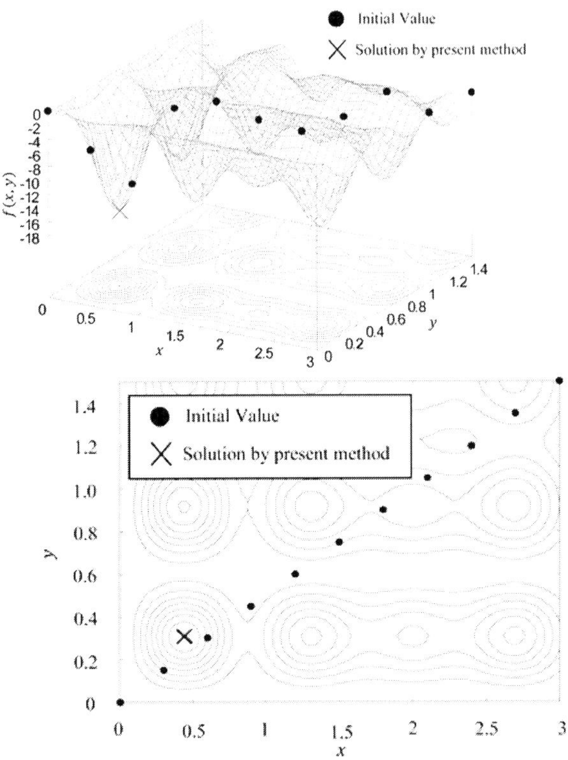

Figure 4. (a) Function space; (b) Contour lines of solution space.

If calculation starts with any initial value marked in the closed dot, the resultant values do not stay at any of the many minimal values in the solution space, but all the values concentrate near the minimum value. Since every solution generated in each calculation is obtained stochastically with this method, the calculation results are not influenced by the initial value in principle. However in reality, the number of calculations is finite. Therefore, the expected value as the minimum value may have a variation. The minimum values in Figure 4(b) have a little variation though it may be invisible. This variation is caused by the limited number of calculations.

Note that it took about 18 seconds to perform one million times of calculations on a 1.4 GHz Pentium 4 machine.

5.2. Line of Swiftest Descent Problem

Next, we applied SPOT to a problem on brachistochrone as one of the classical optimization problems to obtain a function. This problem is demonstrated to obtain a mass point path that shows the brachistochrone between a fixed start point and an end point at different altitudes under the uniform gravitation field. Here, performance function I is time t until the mass point reaches the end point as expressed in the following equation:

$$t = \frac{1}{\sqrt{2g}} \int_0^{y_f} \frac{\sqrt{1+(\mathrm{d}x/\mathrm{d}y)^2}}{\sqrt{y}} \mathrm{d}y \tag{7}$$

The boundary conditions are as follows:

Initial condition: $y(0) = x(0) = 0$, final condition: $y(t_f) = 1.0$, $x(t_f) = \pi/2$.

In this problem, we discretized the vertical axis y at an even division, and generate the values of the horizontal axis x corresponding to y according to the probability distribution of Equation (3) defined by the performance index of Equation (7) to obtain an expected value.

Figure 5 shows the paths obtained with SPOT in the closed dots and the analytic solution in the solid line. It took about 438 seconds on a Pentium 4 machine to perform 20 million times of iterations.

SPOT is to obtain an approximate solution of an optimal solution. However, the approximate solutions agree well with the analytic solutions. Thus, this method can be applied to general optimization problems where performance functions are expressed in functional.

5.3. Optimal Glide Problem of Hang Glider

Lastly, we apply SPOT to a problem of a hang glider that starts the fall and glide at an altitude of 12 m toward the ground to obtain its flight operation (time history of angle of attack) for the maximum down range (as shown in Figure 6).

This problem was firstly solved by Suzuki et al. [10] and their results were used to check the validity of the results of SPOT.

This problem is based on the former problem of [10] and the specifications of the hang glider used here are based on those of the optimal glider obtained in [10] . The shape of the hang glider is shown in Figure 7.

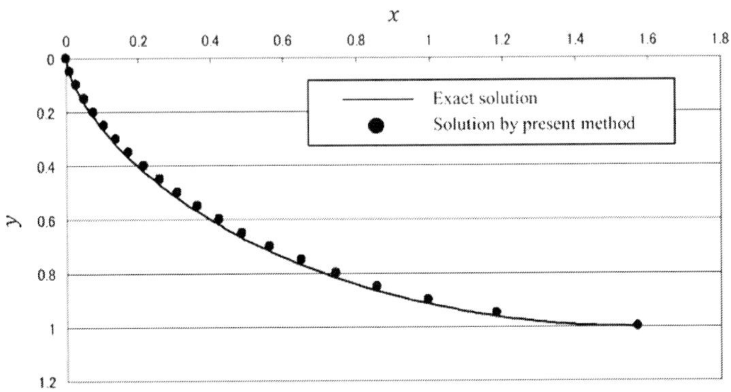

Figure 5. Line of swiftest descent.

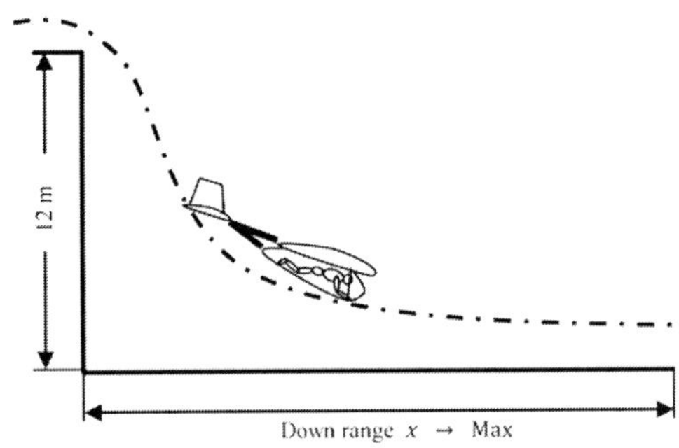

Figure 6. Flight trajectory optimization.

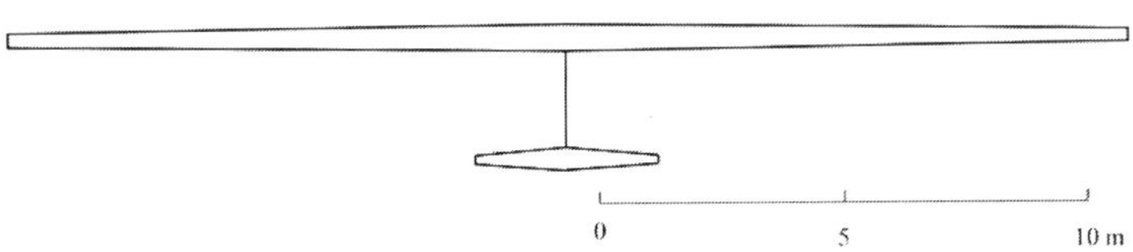

Figure 7. Top view of the hang glider.

Note that the performance index is the reciprocal number of the flight distance to set this problem to a minimal value problem. Also, the maximum value of the load factor is set to the restriction condition. If this condition exceeds the limitation during the flight, a penalty is added to the performance index value.

The equation of motion is as follows:

$$\dot{V} = -(1/2m)\rho V^2 S_1 C_D^* - g\sin\gamma \tag{8}$$

$$\dot{\gamma} = (1/2m)\rho V S_1 C_L^* - (g/V)\cos\gamma \tag{9}$$

$$\dot{x} = V\cos\gamma \tag{10}$$

$$\dot{z} = V\sin\gamma \tag{11}$$

The main conditions are shown as follows:

Performance index:

$$I = 1/x_f + \sum_{n_i > n_{\max}} n_i/x_f \tag{12}$$

Initial condition: $z_0 = 12.0$ m, $V_0 = 5.0$ m/s, $\gamma_0 = -3.5$ deg.

Final condition: $z_f = 2.0$ m.

Restriction condition: $n_{\max} = 3.0$.

Specifications of hang glider: $S_1 = 14.6$ m^2, $A_1 = 36.0$, $b_1 = 22.9$ m, $\lambda_1 = 0.5$, $S_2 = 2.3$ m^2, $A_2 = 6.0$, $\lambda_2 = 0.5$, $m_0 = 35.0$ kg, $m_p = 60.0$ kg.

The performance index value is the reciprocal number of the down range at the point where the hung glider reaches at an altitude of 2 m. First, the control variable, which is the angle of attack in this case, is discretized on the axis of time. Next, values of angle of attack at each time are specified with random numbers. The equations of motion are numerically calculated according to the time history of angle of attack to obtain the flight path. Then, the performance index value is determined as a reciprocal number of the down range. Therefore, the performance index values are obtained when the series of calculations ends with the final conditions being met.

Figure 8(a) and Figure 8(b) show the relation between the down range and the altitude, and that between the down range and the angle of attack, respectively.

This result may be an approximate solution close to the former optimization [9] . Note that it took about 15 hours on a Pentium 4 (1.4 GHz) machine to obtain the calculation result in this example that set the number of iterations to 3 millions.

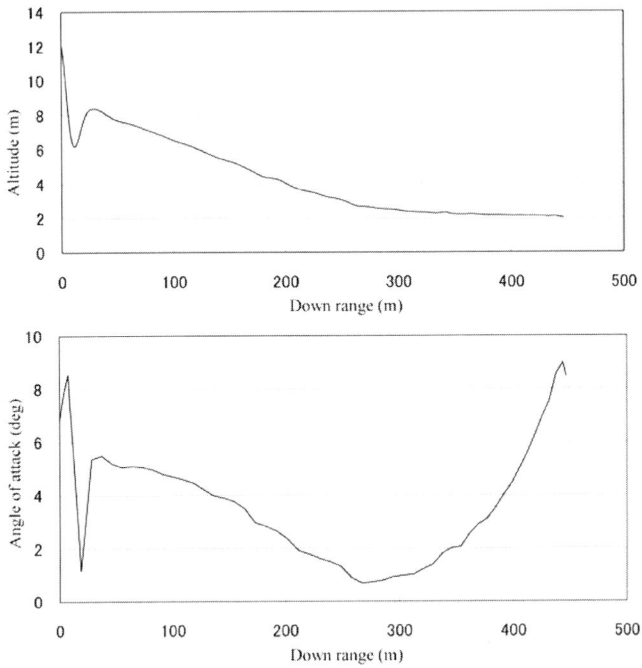

Figure 8. (a) Altitude vs. down range; (b) Angle of attack vs. down range.

6. DISCUSSION

We propose a new optimization method with a stochastic process called SPOT, and then demonstrate numerical calculations.

Generally, when proposing a new method, comparison of performance between the conventional method and a new method should be performed. However, since SPOT is the only method of obtaining the approximate value of the optimal solution, there is no sense in comparing them. Also fair comparison of computation time could not be performed.

As a result, SPOT has the following characteristics:

1) Advantages of Stochastic Process

SPOT results in an expected value from all the solutions generated by calculations. One of the advantages is to avoid influences from the initial value (a start point of the calculation) to the final expected value. The reason is that the generation of solutions during the calculation is a simple stochastic process, and the initial value is only the first-generated solution. As shown in Section 6.1, the calculation results all do not stay at local minimal values even though the calculation starts with any initial value, but the resultant values concentrate near the minimum value. Another advantage is to update an expected value from optimal solutions separately obtained under the same condition. This enables a parallel calculation and the recycling of the former calculation results. This characteristic is unique to SPOT and advantageous to other numerical calculation methods. For example, the calculation of a bivariate function, as shown in Section 6.1, gives the results starting with the multiple initial points as different minimum values. (The cross marks actually show almost the same point.) On the other hand, it is possible to obtain an expected value with all these expected values.

Also, SPOT seems to use asymptotic calculation in the algorithm. However, the former solutions are not used to generate the next solution. This means that the series of calculation is independent and thus the calculation time can easily be estimated with the number of calculations. Therefore, it is easy to estimate the calculation cost.

2) Independency on Experience

Another characteristic is that SPOT has only fluctuation parameter, h, set by operators. There may be a case of extremely low convergence of a solution because of a large fluctuation during calculation with an conventional optimization method using an inclination of the solution curved surface if the features of the solution space are not well known. On the other hand, in principle, the scale of the fluctuation parameter in our method has almost no influence on the process of obtaining an expected value. Although variations of convergence times of calculations due to the scale of the parameter are observed in actual calculations with the limited number of calculations, any large influence on the obtained expected values has not been known. Therefore, this method can be applied to unknown problems, including the surroundings, on solution spaces.

Also, the smaller the fluctuation parameter during calculation, the shorter the calculation time is, though this characteristic is not mentioned in this paper.

3) Restriction Conditions

In the conventional calculation approaches that start with initial solutions to search better performance index values, it may be difficult to generate an initial solution itself and thus to start the numerical calculation if the solution space is narrowed with restriction conditions. However, since the generation of solutions during calculation in our method is a stochastic process, it is possible to perform the calculation regardless of the presence of solutions.

Also, if there are restriction conditions on parameters, which are not the fluctuation parameter: h in (b) but are the load factor: n and the angle of attack: α etc. in Section 6.3 for example, it is necessary to keep parameters within the restriction conditions with general methods. However, with our method, such a device is not necessary. Furthermore, the restriction conditions are rather advantageous to numerical calculations. For example, if there are restrictions on parameter values and the range is specified, solutions can be generated only with values within the restrictions in the process where the series of solutions are generated. This is largely advantageous to perform numerical calculations. The harder the restriction conditions are and the narrower the range of the resultant parameter values is, the more densely the solutions are generated within the restriction conditions and the more easily the calculations are.

Next, in case of restriction conditions on state variables, when a solution generated during calculation is trapped on the restriction, the solution may be abandoned. However, expected values may approach but do not reach the restriction surfaces using only parameter values within constraint ranges and abandoning solutions that break the restrictions. Parameters and solutions that break the restrictions can be applied to a calculation using a penalty method to approach the expected value toward the restriction surface more closely. However, if results with this method are used as initial solutions of deterministic approaches, there is no necessity to make a device.

4) Scope of SPOT

The characteristics of SPOT are to obtain expected values of solution spaces with multiple integral. Thus, there is no problem if the solution spaces have restrictions as mentioned above. However, SPOT is not appropriate to be used for problems that have no restriction on the infinite solution spaces. Also, if a problem has a large solution space and a sharp peak on the performance index value, this method can be applied in principle. However, it may be difficult to obtain an approximate solution with high accuracy in the finite number of calculations. However, it is rare to have infinite range of a variable in an engineering problem, and the upper and lower limits can generally be set. Also, the fluctuation of a performance index value to that of a variable is expected to be smooth. Furthermore, the accuracy required for a variable is finite. Therefore, it is possible to obtain an approximate value for practical use with the finite number of calculations.

7. CONCLUSIONS

We propose a new optimization method based on the path integral called SPOT, and thus perform a numerical calculation on an engineering optimization problem. As a result, we obtain an approximate value of an optimal solution with high accuracy. In addition, this stochastic method has advantages on numerical calculation to other methods.

Also, SPOT has more effective features of optimization than the conventional methods on optimization problems of an unknown system or narrow solution space that leads to difficulty finding a feasible solution.

REFERENCES

1. Kirkpatrick, S., Gelatt Jr., C.D. and Vecchi, M.P. (1983) Optimization by Simulated Annealing. Science, 220, 671-680.

2. Terasaki, M., Kondo, M., Yoshida, H., Yamaguchi, K. and Ishikawa, Y. (2004) Integrated Optimization of Airplane Design and Flight Trajectory by New Optimization Method Using a Stochastic Process. In: CONPUTATIONAL MECHANICS WCCM VI in Conjunction with APCOM'04, Tsinghua University Press & Springer-Verlag, Beijing, 328.

3. Yoshida, H., Konno, T., Yamaguchi, K. and Ishikawa, Y. (2005) New Optimization Method Using Stochastic Process Based on Path Integral. Journal of the Japan Society for Computational Methods in Engineering, 5, 145-150. (in Japanese)

4. Yoshida, H., Yamaguchi, K. and Ishikawa, Y. (2006) Application of a New Optimization Method for System Design. Joint 3rd International Conference on Soft Computing and Intelligent Systems (SCIS) and 7th International Symposium on Advanced Intelligent Systems (ISIS), Tokyo, 20-24 September 2006, 1542-1547.

5. Yoshida, H., Yamaguchi, K. and Ishikawa, Y. (2007) Design Tool Using a New Optimization Method Based on a Stochastic Process. Transactions of the Japan Society for Aeronautical and Space Sciences, 49, 220-230.

6. Feynman, R.P. and Hibbs, A.R. (1965) Quantum Mechanics and Path Integrals. McGraw-Hill, New York.

7. Metropolis, N., Rosenbluth, A.W., Rosenbluth, M.N., Teller, A.H. and Teller, E. (1953) Equation of State Calculations by Fast Computing Machines. Journal of Chemical Physics, 21, 1087-1092.

8. Nakane, M., Kobayashi, D., Yoshida, H. and Ishikawa, Y. (2009) Feasibility Study on Single Stage to Orbit Space Plane with RBCC Engine. 16th AIAA/DLR/DGLR International Space Planes and hypersonic Systems and Technologies Conference, Bremen, AIAA 2009-7331.

9. Ishimori, Y., Nakane, M., Ishikawa, Y., Yoshida, H. and Yamaguchi, K. (2011) Feasibility Study for Spaceplane's Concepts in Ascent Phase Using the Optimization Method. Journal of the Japan Society for Aeronautical and Space Sciences, 59, 291-297. (in Japanese)

10. Suzuki, S. and Kawamura, N. (1996) Simultaneous Optimization of Sailplane Design and Its Flight Trajectory. Journal of Aircraft, 33, 567-571.

Chapter 5

A New Formula for Partitions in a Set of Entities into Empty and Nonempty Subsets, and Its Application to Stochastic and Agent-Based Computational Models

Ghennadii Gubceac, Roman Gutu, Florentin Paladi

Department of Theoretical Physics, State University of Moldova, Chisinau, Republic of Moldova

ABSTRACT

In combinatorics, a Stirling number of the second kind $S(n,k)$ is the number of ways to partition a set of n objects into k nonempty subsets. The empty subsets are also added in the models presented in the article in order to describe properly the absence of the corresponding type i of state in the system, i.e. when its "share" $p_i = 0$. Accordingly, a new equation for partitions $P(N,m)$ in a set of entities into both empty and nonempty subsets was derived. The indistinguishableness of particles (N identical atoms or molecules) makes only sense within a cluster (subset) with the size $0 \le n_i \le N$. The first-order phase transition is indeed the case of transitions, for example in the simplest interpretation, from completely liquid state $\text{type}L = \{n_1 = N, n_2 = 0\}$ to the completely crystalline state $\text{type}C = \{n_1 = 0, n_2 = N\}$. These partitions are well distinguished from the physical point of view, so they are 'typed' differently in the model. Finally, the present developments in the physics of complex systems, in particular the structural relaxation of supercooled liquids and glasses, are discussed by using such stochastic cluster-based models.

KEYWORDS

Partitions; Agent-Based Models; Stochastic Processes; Complex Systems

1. INTRODUCTION

In mathematics, particularly in combinatorics and the study of partitions, a Stirling numbers of the second kind $S(n,k)=\left\{ n \atop k \right\}$ count the number of ways to partition a set of n labeled objects into k nonempty unlabelled subsets. By the way, Stirling numbers of the second kind show up more often than those of first and third kind (or Lah numbers), and James Stirling himself considered this kind first [1]. Equivalently, they count the number of different equivalence relations with precisely k equivalence classes that can be defined on an n element set, and they can be calculated using the explicit formula $S(n,k)=\frac{1}{k!}\sum_{j=0}^{k}(-1)^{k-j}\binom{k}{j}j^n$, where $\binom{k}{j}$ is the binomial coefficient. The sum over the values for *k* of the Stirling numbers of the second kind, i.e. $\sum_{k=0}^{n}\left\{ n \atop k \right\}=B_n$ gives the nth Bell number, that is the total number of partitions of a cluster with n entities or agents [2]. The question arises when the empty subsets or clusters are needed to be used in the models, for example, with heterogeneous structure interactions and, subsequently, the partition by Stirling numbers of the second kind becomes inappropriate to count partitions in such complex systems [3].

In general, agent-based modeling is currently a technique widely used to simulate complex systems in computer science and social sciences. On the other hand, a Markovian process is a stochastic process whose future probabilities are determined by its most recent values. The agent-based computational models (ABM) fits well this description, except for the cases when decisions are dependent on the state of the systems of more than one steps ago, which is the case when ABM agents experience learning, adaptation, and reproduction [4].

In this study, a general equation that describes clustering process among interacting agents in heterogeneous populations, i.e. the partition process in a set of entities into empty and nonempty clusters, is derived and used to study how different behavioral norms affect the individual and social welfare in a population with heterogeneous preferences. One can consider them as an idealization of an imperative and a more liberal approach to social norms or stylized behavioral rules studied by agent-based computational models [5,6]. Another application refers to the generic stochastic model for crystal nucleation which is useful to depict the impact of interface between the nucleus considered as a cluster of a certain number of molecules and the liquid phase for the enhancement of the overall nucleation process. It is generally known that first-order phase transitions occur by nucleation mechanism, and both the nucleus, a cluster of atoms or molecules, and the nucleation work, a energy barrier to the phase transition, are basic thermodynamic quantities in the theory of nucleation. However, the critical nucleus formation is statistically a random event with a probability largely determined by the nucleation work which increases with the subnuclei size [7]. The traditional differential equation modeling is not the alternative to agent-based models; only a set of differential equations, each describing the dynamics of one of the system's constituent units, is an agent-based model [8].

The general formulation is outlined in Section 2. In Section 3 a probabilistic approach to the crystal nucleation process is considered. The main conclusions are presented in Section 4.

2. THE MODEL

There are N entities which can be in 3 different states (call them left, center and right), and can play 3 actions (again left, center and right). Interaction in this agentbased model involves always one active and one passive player, but agents can play both roles interchangeably. They have preferences over their states: accept one state, are neutral with respect to another state and reject the remaining state. When two agents meet, the active player sets the passive player's state according to his actionwhich in turn is determined by one of the applied rule. This identifies only 6 possible combinations. Denote with $p_1 \cdots p_6$ the shares of the population characterized by each combination of preferences, as in Table 1. That is, drawing randomly one agent, it will be of type i with probability p_i. After each interaction, the passive player gets a payoff of +1 if it is in the accepted state, a payoff of 0 if it is in the neutral state, and a payoff of −1 if it is in the rejected state. The active player does not get any feedback. If the active player follows the first J-rule, it always plays the action corresponding to the accepted state. If it follows the second H-rule, it randomizes between actions corresponding to the accepted and neutral states. The further example will clarify. Suppose two individuals, A and B, meet. Player

A is the active one, and rejects left, accepts right, and is thus neutral with respect to the center. Player B is the passive one. It accepts left, rejects right, and is neutral with respect to center, like player A. Suppose A follows the J-rule, and will play right, setting B's state to right. B will then have a payoff of −1. Suppose, on the other hand, that A follows the H-rule, and will randomize between center and right. The payoff for B could then be either 0 or −1. Note that there is no strategic interaction in the model: the passive player's payoff depends on the active player's choice, but the active player's choice does not depend on the passive player in any way, so the game-theoretic solution concepts like Nash equilibrium become useless.

Aggregate results are defined in terms of both the mean π and the variance σ^2 of the payoffs which denote the stability and the heterogeneity of population, respectively. However, in order to avoid arbitrary choices we do not specify a particular functional form, and report separately the results for the mean and the variance. Let $N = 1, 2, \cdots, \infty$ be the total number of entities in the model, and the $\{n_1, n_2, n_3, n_4, n_5, n_6\}$ is their partition into $m = 6$ subsets. Each subset can be called cluster, and the process itself—clustering. The size of each cluster can vary from 0 to N, $n_i = \overline{0, N}, i = \overline{1, 6}$, and $\sum_{i=1}^{6} n_i = N$. The number of possible partitions P is a function of N and m, and the explicit solution is

$$
\begin{aligned}
&P(N, m=6) \\
&= {}_6C_1 + (N-1)\,{}_6C_2 + \left(1 + \sum_{i=2}^{N-2} i\right){}_6C_3 \\
&+ \left(N - 3 + (N-2)\sum_{i=2}^{N-3} i - \sum_{i=2}^{N-3} i^2\right){}_6C_4 \\
&+ \left(1 + \sum_{i=2}^{N-4} i + \sum_{i=3}^{N-3} i \sum_{l=2}^{i-1} l - (N-3)\sum_{i=2}^{N-4} i^2 + \sum_{i=2}^{N-4} i^3\right){}_6C_5 \\
&+ (N-5)\sum_{i=2}^{N-5} (N-i-4)i + \sum_{i=3}^{N-4} (N-i-3)i \sum_{l=2}^{i-1} l \\
&- \sum_{i=2}^{N-5} i^2 \sum_{l=i+1}^{N-4} l + \left(\sum_{i=2}^{N-5} (N-i-4)i^3\right){}_6C_6 \\
&= 2(74N - 213) + 2(11N - 13)\sum_{i=2}^{N-5} i \\
&+ 2(1-3N)\sum_{i=2}^{N-5} i^2 + (2+N)\sum_{i=2}^{N-5} i^3 - \sum_{i=2}^{N-5} i^4 \\
&+ \sum_{i=3}^{N-4} \left((3+N)i - i^2\right)\sum_{l=2}^{i-1} l - \sum_{i=2}^{N-5} i^2 \sum_{l=i+1}^{N-4} l \\
&= \frac{1}{5!}\prod_{i=1}^{5} (N+i),
\end{aligned}
$$

Table 1. Distribution of preferences in the heterogeneous system with 6 types of states.

Type	Accepted State	Rejected State	Share	*left*	*center*	*right*
1	*left*	*center*	p_1	p_1	$-p_1$	0
2	*left*	*right*	p_2	p_2	0	$-p_2$
3	*center*	*left*	p_3	$-p_3$	p_3	0
4	*center*	*right*	p_4	0	p_4	$-p_4$
5	*right*	*left*	p_5	$-p_5$	0	p_5
6	*right*	*center*	p_6	0	$-p_6$	p_6
	Coordinates			$p_1 + p_2 - p_3 - p_5$	$-p_1 + p_3 + p_4 - p_6$	$-p_2 - p_4 + p_5 + p_6$

where ${}_6C_1 = {}_6C_5 = 6, {}_6C_2 = {}_6C_4 = 15, {}_6C_3 = 20, {}_6C_6 = 1$ are combinations given by the binomial coefficient ${}_mC_k = \binom{m}{k} = \frac{m!}{(m-k)!k!}$. In general, when the number of subsets $m > 1$, the number of possible partitions P of N agents into m subgroups or subsets is

$$P(N,m) = \frac{1}{(m-1)!}\prod_{i=1}^{m-1}(N+i) \quad . (1)$$

The matrix is diagonally symmetric

$$P(N,i+1) = P(i,N+1)$$

for $i = 0,1,\cdots,N-1$, and it can be formed by arranging the partition numbers according to the parameters N and m. Table 2 contains this array of values for the given numbers of partitions. The reccurence relation is

$$P(N,m) = P(N-1,m) + P(N,m-1)$$

for $m > 0$ with the initial condition $P(0,m>0) = 1$. For instance, the number 330 in column $m = 5$ and row $N = 7$ is given by $330 = 210 + 120$, where 210 is the number above and 120 is the number on the left of 330. The diagonal elements of the bi-triangular array of values for the numbers of partitions are 1, 2, 6, 20, 70, 252, 924, 3432, 12,870, 48,620, ..., and it is represented by nth central binomial coefficient:

$$C(2n,n) \equiv \binom{2n}{n} = \frac{(2n)!}{(n!)^2}$$

for all $n \geq 0$.

They are called central since they show up exactly in the middle of the even-numbered rows in Pascal's traingle. These numbers have the generating function

$$\frac{1}{\sqrt{1-4x}} = 1 + 2x + 6x^2 + 20x^3 + 70x^4 + 252x^5$$
$$+924x^6 + 3432x^7 + \cdots.$$

It is known that the asymptotics for the central binomial coefficient $C(2n,n)$ can be written in the form of a particular case of the Wallis formula, i.e.

$$\lim_{n\to\infty}\frac{2^{4n}}{n\binom{2n}{n}^2}=\pi\lim_{n\to\infty}\frac{n\left[\Gamma(n)\right]^2}{\left[\Gamma\left(\frac{1}{2}+n\right)\right]^2}=\pi$$
,

Table 2. Bi-triangular array of values for the numbers of partitions P(N,m).

N,m	1	2	3	4	5	6	7	8	9	10	Recurrence Relation
0	1	1	1	1	1	1	1	1	1	1	
1	1	2	3	4	5	6	7	8	9	10	
2	1	3	6	10	15	21	28	36	45	55	
3	1	4	10	20	35	56	84	120	165	220	
4	1	5	15	35	70	126	210	330	495	715	
5	1	6	21	56	126	252	462	792	1287	2002	$P(N,m) = P(N-1,m) + P(N,m-1)$
6	1	7	28	84	210	462	924	1716	3003	5005	
7	1	8	36	120	330	792	1716	3432	6435	11,440	
8	1	9	45	165	495	1287	3003	6435	12,870	24,310	
9	1	10	55	220	715	2002	5005	11,440	24,310	48,620	
10	1	11	66	286	1001	3003	8008	19,448	43,758	92,378	

where $\Gamma(x)$ is the gamma function, so

$$\binom{2n}{n}\sim\frac{4^n}{\sqrt{\pi n}},\quad n\to\infty$$
.

By the way, this equation can also be used to determine the constant $\sqrt{2\pi}$ in front of the Stirling's formula.

Now it is straightforward to see that when all individuals share the same preferences (polarization) the first rule gives a higher payoff. In the other extreme case, when preferences are equally distributed in the ensemble (dispersion) and $p_1=p_2=\cdots=p_6=\frac{1}{6}$, it is again straightforward to see that the rules are equivalent, and lead to an average payoff $\pi=0$. Consider next an active player of type 1 (accepts left and rejects center) which meets in turn all other (passive) agents, including himself. If it follows the first rule, then it will play left causing a payoff of +1 in $(p_1+p_2)N$ agents, and a payoff of −1 in $(p_3+p_5)N$ agents. Note that there are $(p_1+p_2)N$ similar entities in the ensemble. Suppose now that everybody meets everybody else both as active and as passive players. Coupling them randomly and randomly choosing active and passive players only adds some noise to the simulation results. So the average payoff when everybody plays according to the first J-rule is

$$\pi_1 = (p_1 + p_2)(p_1 + p_2 - p_3 - p_5) + (p_3 + p_4)(-p_1 + p_3 + p_4 - p_6) + (p_5 + p_6)(-p_2 - p_4 + p_5 + p_6). \quad (2)$$

Similarly, the average payoff with the second H-rule is

$$\pi_2 = \frac{1}{2}(p_1(p_1 - p_3 - p_5 + p_6) + p_2(p_2 - p_3 + p_4 - p_5) + p_3(-p_1 + p_3 + p_5 - p_6) + p_4(-p_1 + p_2 + p_4 - p_6) + p_5(-p_2 + p_3 - p_4 + p_5) + p_6(p_1 - p_2 - p_4 + p_6)). \quad (3)$$

To study the behavior of $\pi_1 - \pi_2$ based on Equations (2) and (3), it is convenient to set some of the probabilities to zero. It is straightforward to see that when there is just one probability different from 0 (and thus equal to 1) we have $\pi_1 - \pi_2 = \frac{1}{2} > 0$. In general, the number of different possible combinations of non-zero probabilities is given by the above binomial coefficient $_mC_k$.

Thus for two non-zero probabilities we obtain a set of $_6C_2 = 15$ equations which can be grouped in just 3 different functional forms shown in Table 3. Figure 1 graphs these 3 curves for all values of p_i and $p_j = 1 - p_i$. The J-rule still performs better in all cases but one, when the two rules are equivalent. However, from three nonzero probabilities onward things start to look differently. For three non-zero probabilities we have $_6C_3 = 20$ equations, while for four non-zero probabilities there are $_6C_4 = 15$ equations, and for five non-zero probabilities $_6C_5 = 6$ equations. These equations reduce to just three different functional forms in case of three and four nonzero probabilities, and to just one expression in case of five non-zero probabilities.

Table 3. Expressions of $\pi_1 - \pi_2$ for different numbers of non-zero probabilities p_i, i = 1, 2, ⋯, 6.

No. p_i	Non-zero Probabilities	Expressions for $\delta = \pi_1 - \pi_2$
	(p_1, p_2) (p_3, p_4) (p_5, p_6)	$\delta_1 = \frac{1}{2} + p_i - p_i^2$
2	(p_1, p_3) (p_1, p_6) (p_2, p_4) (p_2, p_5) (p_3, p_5) (p_4, p_6)	$\delta_2 = \frac{1}{2}(1 - 2p_i)^2$
	(p_1, p_4) (p_1, p_5) (p_2, p_3) (p_2, p_6) (p_3, p_6) (p_4, p_5)	$\delta_3 = \frac{1}{2}(1 + 3p_i(p_i - 1))$
	(p_1, p_2, p_3) (p_1, p_2, p_4) (p_1, p_2, p_5) (p_1, p_2, p_6) (p_1, p_3, p_4) (p_2, p_3, p_4) (p_3, p_4, p_5) (p_3, p_4, p_6) (p_1, p_5, p_6) (p_2, p_5, p_6) (p_3, p_5, p_6) (p_4, p_5, p_6)	$\delta_1 = \frac{1}{2}(p_h^2 + p_i^2 + p_j^2 + 4p_h p_i - 2p_h p_j - p_i p_j)$
3	(p_1, p_3, p_5) (p_1, p_3, p_6) (p_2, p_3, p_5) (p_2, p_4, p_5) (p_1, p_4, p_6) (p_2, p_4, p_6)	$\delta_2 = \frac{1}{2}(p_h^2 + p_i^2 + p_j^2 - 2(p_h p_i + p_i p_j) - p_h p_j)$
	(p_2, p_3, p_6) (p_1, p_4, p_5)	$\delta_3 = \frac{1}{2}(p_h^2 + p_i^2 + p_j^2 - p_h p_i - p_h p_j - p_i p_j)$
	(p_1, p_2, p_3, p_4) (p_1, p_2, p_5, p_6) (p_3, p_4, p_5, p_6)	$\delta_1 = \frac{1}{2}(p_h^2 + p_i^2 + p_j^2 + p_k^2 + 4(p_h p_i + p_j p_k) - 2(p_h p_j + p_i p_k) - p_i p_j - p_h p_k)$
4	(p_1, p_2, p_3, p_5) (p_1, p_2, p_4, p_6) (p_1, p_3, p_4, p_6) (p_1, p_3, p_5, p_6) (p_2, p_3, p_4, p_5) (p_2, p_4, p_5, p_6)	$\delta_2 = \frac{1}{2}(p_h^2 + p_i^2 + p_j^2 + p_k^2 + 4p_h p_i - 2(p_h p_j + p_i p_k + p_j p_k) - p_i p_j - p_i p_k)$
	(p_1, p_2, p_3, p_6) (p_1, p_2, p_4, p_5) (p_1, p_3, p_4, p_5) (p_2, p_3, p_4, p_6) (p_2, p_3, p_5, p_6) (p_1, p_4, p_5, p_6)	$\delta_3 = \frac{1}{2}(p_h^2 + p_i^2 + p_j^2 + p_k^2 + 4p_h p_i - 2(p_h p_j + p_h p_k) - p_i p_j - p_i p_k - p_j p_k)$
5	$\delta = \frac{1}{2}(p_h^2 + p_i^2 + p_j^2 + p_k^2 + p_l^2 + 4(p_h p_i + p_j p_k) - 2(p_h p_j + p_i p_k + p_i p_l + p_j p_l) - p_i p_j - p_h p_k - p_h p_l - p_k p_l)$	

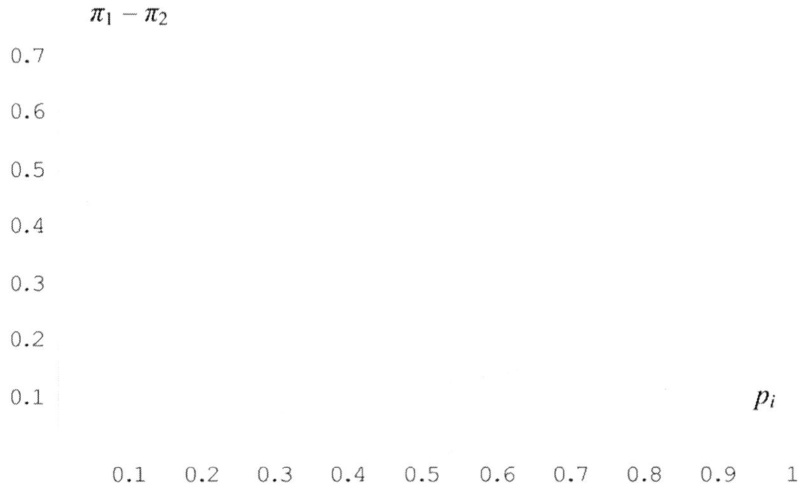

Figure 1. $\pi_1 - \pi_2$ in the case of two non-zero probabilities, p_i and $p_j = 1 - p_i$.

A trend towards a better performance of the H-rule, as the distribution of preferences in the system becomes less polarized, is evident. However, in order to better investigate it, a definition of how much preferences are polarized is needed. We represent the distribution of states as a single point in a three dimensional space, where the axes are labeled l, c and r. The l coordinate is found by counting all agents who accept left, and subtracting all agents who reject left. The result is then normalized to the size of the population. Similarly for the other two coordinates. Hence,

$$
\begin{aligned}
l &= p_1 + p_2 - p_3 - p_5, \\
c &= p_3 + p_4 - p_1 - p_6, \\
r &= p_5 + p_6 - p_2 - p_4,
\end{aligned}
\tag{4}
$$

where $l + c + r = 0$. Note that different distributions of states can lead to the same point in the sphere. For instance, the point in the origin is given not only by $p_1 = p_2 = \cdots = p_6 = 1/6$, but by any combination of preferences such as $p_1 = p_3$, $p_2 = p_5$, and $p_4 = p_6$. Taking into account Equation (4), one can define now the polarization of states as the distance from the center of the sphere:

$$
d(l,r,c) \equiv d(p_1, p_2, \cdots, p_6) = \sqrt{l^2 + r^2 + c^2}. \tag{5}
$$

Note that $d \in \left[0, \sqrt{2}\right]$: all points thus lie inside a sphere around the origin.

The variances σ_1^2 and σ_2^2 are defined for each discrete distribution $D \equiv 1,2$ with the expectation (mean) value π_D as follows:

$$\sigma_D^2 = \sum_{i=1}^{6} p_i \left(\pi_{i,D} - \pi_D\right)^2, \tag{6.1}$$

where

$$\begin{aligned}
\pi_{1,1} &= p_1 + p_2 - p_3 - p_4, \\
\pi_{4,1} &= p_3 + p_4 - p_5 - p_6, \\
\pi_{2,1} &= p_1 + p_2 - p_5 - p_6, \\
\pi_{5,1} &= -p_1 - p_2 + p_5 + p_6, \\
\pi_{3,1} &= -p_1 - p_2 + p_3 + p_4, \\
\pi_{6,1} &= -p_3 - p_4 + p_5 + p_6
\end{aligned} \tag{6.2}$$

and

$$\begin{aligned}
\pi_{1,2} &= \left(p_1 - p_3 - p_5 + p_6\right)/2, \\
\pi_{4,2} &= \left(-p_1 + p_2 + p_4 - p_6\right)/2, \\
\pi_{2,2} &= \left(p_2 - p_3 + p_4 - p_5\right)/2, \\
\pi_{5,2} &= \left(-p_2 + p_3 - p_4 + p_5\right)/2, \\
\pi_{3,2} &= \left(-p_1 + p_3 + p_5 - p_6\right)/2, \\
\pi_{6,2} &= \left(p_1 - p_2 - p_4 + p_6\right)/2.
\end{aligned} \tag{6.3}$$

Figure 2 explores how outcomes vary as functions of the distance $d(l,r,c)$ defined by Equation (5). The whole range $[0,1]$ is sampled, for all probabilities $p_1 \cdots p_6$. The step considered for creating all combinations of probabilities is 0.025, i.e. the total number of agents is 40. The average values for π_1 and π_2 are shown in Figure 2(a), and for each value of the distance from the center of the sphere, $d(l,r,c)$, the frequency of wins with each rule is computed (see Figure 2(b)). When $\pi_1 - \pi_2 > 0$ a win is assigned to the first J-rule, and when $\pi_1 - \pi_2 < 0$ a win is assigned to the second one. Exactly in the center of the sphere the two rules lead to the same payoff, independently of the underlying distribution of states. Close to the center, each rule wins in about 50% of the cases. Then, as we move away from the center, the first rule improves its performance, and it is always better when the states are totally polarized, but the total number of states for intermediate values of the distance d is much larger than for the dispersed and polarized states.

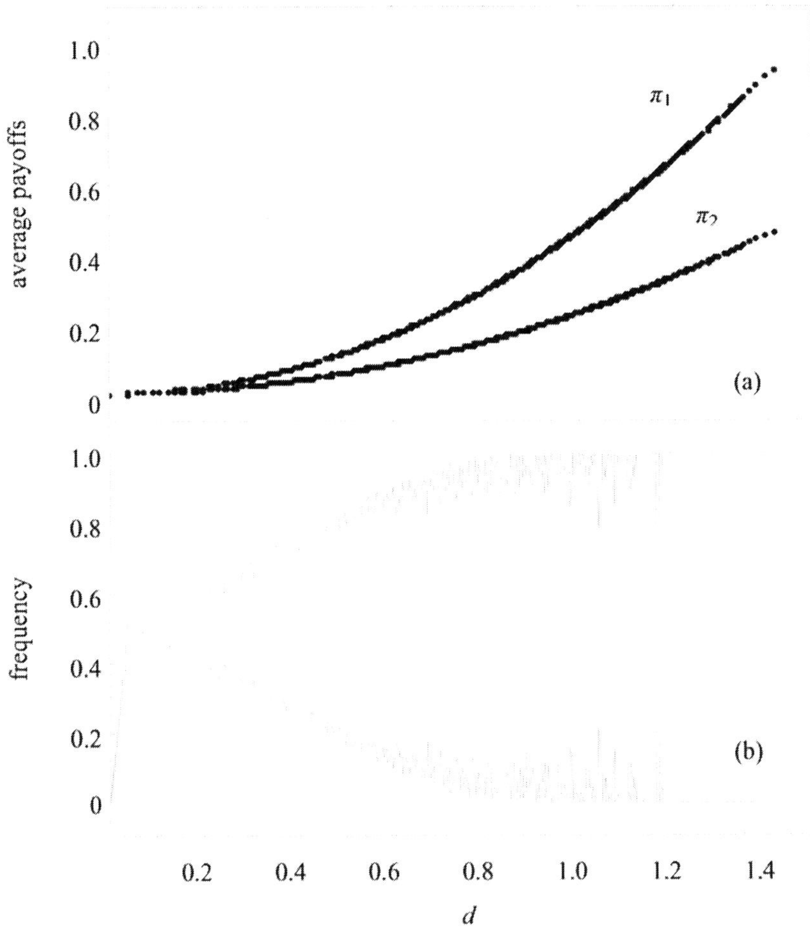

Figure 2. Average values for π_1 (upper line) and π_2 (a), and relative frequency of negative and positive values for $\pi_1 - \pi_2$ (b) as functions of the distance $d(p_1, p_2, \cdots, p_6)$ from the center of sphere with a radius of $\sqrt{2}$: 40 entities, 6 clusters in a similar ABM model.

Meanwhile, the variance (6.1) with the J-rule (6.2) is generally higher than the variance with the H-rule (6.3), especially when the preferences are dispersed in the system, and when they are quite polarized. On average, however, when one rule is better in terms of higher expected payoffs it is also better in terms of lower heterogeneity, and there are the same average values of the scale-free coefficients of variation on the distance d for any level of fragmentation of states.

3. A PROBABILISTIC APPROACH TO THE CRYSTAL NUCLEATION PROCESS

The second application refers to the nucleation process, a widely spread phenomenon in both nature and technology, which may be considered as a representative of the aggregation phenomena in complex systems. Let's consider N atoms which can be in 3 different states (cluster, liquid and their interface), and can perform 4 possible moves: liquid to interface, interface to liquid, interface to cluster, and cluster to interface. One can identify 4 different combinations denoted with probabilities $p_1 \cdots p_4$, as in Table 4. That is, drawing randomly one particle, it will be of type i with probability p_i Let $N = 1, 2, \cdots, \infty$ be the total number of atoms in the system, and $\{n_1, n_2, n_3, n_4\}$ are their partition into 4 subsets. Each subset can be called cluster, and the number of possible partitions (1) in this case is

$$P(N, m=4) = \frac{1}{3!}\prod_{i=1}^{3}(N+i),$$

where $n_i = \overline{0, N}, i = \overline{1, 4}$ and $\sum_{i=1}^{4} n_i = N$. For example, in a system of N=1000 atoms, $P(N, m=4)$ equals to 167,668,501! Accordingly, the number of repeated computer runs in an ABM model would be very large due to different possible partitions. But we are able to overcome this problem, as already mentioned in the previous section, by developing similar stochastic mathematical models which can describe exactly the results of the agentbased computational models, and, finally, by bridging the gap between ABM modeling and stochastic processes.

Let's consider further that each particle interacts with the entire group both as an aggressor in terms of the Kolmogorov mathematical theory [9], and as a passive agent in terms of the ABM computational models as well. Then the mean π, namely the stability index, takes here the form

$$\pi = p_1(p_1 - p_3) + (p_2 + p_3)(-p_1 + p_2 + p_3 - p_4) + p_4(p_4 - p_2) \quad (7.1)$$

or, taking into account that $\sum_{i=1}^{4} p_i = 1$, one can exclude one probability, for example p_4, from the above equation:

$$\pi = p_1(p_1 - p_3) + (1 - p_1 - 2p_2 - p_3)(1 - p_1 - p_2 - p_3) + (p_2 + p_3)(-1 + 2p_2 + 2p_3). \quad (7.2)$$

One can represent again the distribution of states as a three-dimensional point (l,r,c) inside a sphere, and the mean payoff π (Equation (7.1) or Equation (7.2)) can be obtained as a function of the distance

$$d(l,r,c) \equiv d(p_1, p_2, p_3, p_4) = \sqrt{l^2 + r^2 + c^2}$$

where the axes are labeled l, c and $r: l = p_1 - p_3$, $c = p_2 + p_3 - p_1 - p_4, r = p_4 - p_2$, and, as stated in the previous section, $l + c + r = 0$ and $d \in [0, \sqrt{2}]$. Thus different distributions of states can lead to the same point in the sphere, i.e. different microscopic partitions can generate the same result on aggregate inside a sphere around the origin. The results for the two limit cases are obviously: if all particles would show the same behavior, then $d = \sqrt{2}$ and there is a maximum stability of states in such a completely asymmetrical system, but $\pi = 0$ for a homogeneous system, $p_1 = p_2 = p_3 = p_4 = 1/4$, and for combinations such as $p_1 = p_3$ and $p_2 = p_4$ in the case of unstable states.

Table 4. Distribution of accepted and rejected states in the system.

Type	Accepted State	Rejected State	Share	*cluster*	*interface*	*liquid*
1	*cluster/nucleus*	*interface*	p_1	p_1	$-p_1$	0
2	*interface*	*liquid*	p_2	0	p_2	$-p_2$
3	*interface*	*cluster/nucleus*	p_3	$-p_3$	p_3	0
4	*liquid*	*interface*	p_4	0	$-p_4$	p_4
Coordinates				$p_1 - p_3$	$-p_1 + p_2 + p_3 - p_4$	$-p_2 + p_4$

4. CONCLUSIONS

Mathematical models developed in this paper were used together with agent-based computations to study interactions in two quite different heterogeneous systems. The models are general ones and allow simulations for any size of the system. Our results also support the idea that general properties observed in the complex systems of any kind, which arise from the interplay between random interactions and their complex structures, could be successfully investigated by the same tool. For example, these properties are frequently found in the physical sciences, and are the domain of non-equilibrium statistical mechanics. Because of the real limitations in the use of analytical methods to study such problems, it is often necessary to resort to numerical ones, and the advent of computer simulations has led to an increase of scientific activity in this area that has emerged nowadays as a major subject of interdisciplinary research.

The recent discussions about the gap between agentbased computational models and stochastic analytical models have stimulated the research on this topic [3,10]. In such context, the article supports a unified probabilistic approach to simulation of multi-agent interactions in heterogeneous complex systems. This method appears phenomenological if it is not agent-based, and we prove analytically that the average outcomes of multiple agentbased computations can be described precisely. Furthermore, a useful aggregation procedure for representing the three-dimensional distribution of states, and a general formula that describes clustering process among interacting agents in heterogeneous populations, which is the Equation (1) for partitions in a set of entities into empty and nonempty subsets, are proposed. Bi-triangular array of values for the numbers of partitions is presented, and its diagonal elements are represented by nth central binomial coefficient.

In particular, we obtained that different distributions of states can lead to the same point inside the sphere around the origin, i.e. different microscopic partitions can generate the same aggregate result. It would be a particular conclusion set for the analyzed situations, but one would expect this in different circumstances due to the universal aspects observed in the behavior of complex systems.

ACKNOWLEDGEMENTS

F.P. thanks Dr. Matteo Richiardi for his ABM contribution.

REFERENCES

1. R. L. Graham, D. E. Knuth and O. Patashnik, "Concrete Mathematics. A Foundation for Computer Science," 2nd Edition, Addison-Wesley Professional, Reading, 1994.
2. J. Sandor and B. Crstici, "Handbook of Number Theory II," Kluwer Academic Publishers, Dordrecht, 2004.
3. F. Paladi, "On the Probabilistic Approach to Heterogeneous Structure Interactions in Agent-Based Computational Models," Applied Mathematics and Computation, Vol. 219, No. 24, 2013, pp. 11430-11437.
4. E. Bonabeau, "Agent-Based Modeling: Methods and Techniques for Simulating Human Systems," Proceedings of the National Academy of Sciences of the United States of America, Vol. 99, No. 3, 2002, pp. 7280-7287.
5. M. Richiardi and F. Paladi, "Jesus, Hillel and the Man of the Street. Moral and Social Norms in Heterogeneous Populations," LABORatorio R. Revelli Working Paper 40, 2005. http://eco83.econ.unito.it/terna/swarmfest2005papers/richiardi_paladi.pdf
6. M. Richiardi, "Jesus vs Hillel. From Moral to Social Norms and Back," European Journal of Economic and Social Systems, Vol. 19, No. 2, 2006, pp. 171-190.
7. D. Kashchiev, "Nucleation. Basic Theory with Applications," Butterworth-Heinemann, Oxford, 2000.
8. C.W. Gardiner, "Handbook of Stochastic Methods: for Physics, Chemistry and the Natural Sciences," 2nd Edition, Springer-Verlag, Berlin, 1985.
9. A. N. Kolmogorov, "On Statistical Theory of Metal Crystallisation (in Russian)," Izvestiya Academy of Sciences, USSR, Mathematics, Vol. 3, 1937, pp. 355-360.
10. L. Feng, B. Li, B. Podobnik, T. Preis and H. E. Stanley, "Linking Agent-Based Models and Stochastic Models of Financial Markets," Proceedings of the National Academy of Sciences of the United States of America, Vol. 109, No. 22, 2012, pp. 8388-8393.

Chapter 6

Stochastic Approximation Method for Fixed Point Problems

Ya. I. Alber1, C. E. Chidume2, Jinlu Li3

1Department of Mathematics, The Technion-Israel Institute of Technology, Haifa, Israel

2The Abdus Salam International Centre for Theoretical Physics, Trieste, Italy

3Department of Mathematical Sciences, Shawnee State University, Portsmouth, USA

ABSTRACT

We study iterative processes of stochastic approximation for finding fixed points of weakly contractive and nonexpansive operators in Hilbert spaces under the condition that operators are given with random errors. We prove mean square convergence and convergence almost sure (a.s.) of iterative approximations and establish both asymptotic and nonasymptotic estimates of the convergence rate in degenerate and non-degenerate cases. Previously the stochastic approximation algorithms were studied mainly for optimization problems.

KEYWORDS

Hilbert Spaces; Stochastic Approximation Algorithm; Weakly Contractive Operators; Nonexpansive Operators; Fixed Points; Convergence in Mean Square; Convergence Almost Sure (a.s.); Nonasymptotic Estimates of Convergence Rate.

1. INTRODUCTION

In this paper the following problem is solved: To find a fixed point x^* of the operator $T:G \to H,$ in other words, to find a solution $x^* \in G$ of the equation

$$x = Tx, \tag{1.1}$$

where T is a Lipschitz continuous mapping, H is a Hilbert space, $G \subseteq H$ is a closed convex subset. We suppose that x^* exists, i.e., the fixed point set N of T is nonempty. Note in different particular cases of the Equation (1.1), for example, when $T:G \to G,$ the solution existence and solution uniqueness can be proved under some additional assumptions.

We separately consider two classes of mappings T: the class of weakly contractive maps and more general class of nonexpansive ones. Let us recall their definitions.

Definition 1.1. A mapping $T:G \to H$ is said to be weakly contractive of class $C_{\phi(t)}$ on a closed convex subset $G \subset H$ if there exists a continuous and increasing function $\phi(t)$ defined on IR^+ such that ϕ is positive on $IR^+/\{0\}, \phi(0) = 0, \lim_{t \to +\infty}\phi(t) = +\infty,$ and for all

$$x, y \in G, \|Tx - Ty\| \le \|x - y\| - \phi(\|x - y\|). \tag{1.2}$$

Remark 1.2. It follows from (1.2) that $\psi(t) \le t$ and in real problems an argument t of the function $\psi(t)$ doesn't necessary approaches to ∞ obeying the condition $T:G \to H$ (see the example in Remark 3.4).

Definition 1.3. A mapping $T:G \to H$ is said to be nonexpansive on the closed convex subset $G \subset H$ if for all $x, y \in G$

$$\|Tx - Ty\| \le \|x - y\|.$$

It is obvious that the class of weakly contractive maps is contained in the class of nonexpansive maps because the right-hand side of (1.2) is estimated as

$$0 \le \|x - y\| - \phi(\|x - y\|) \le \|x - y\|, \tag{1.3}$$

and it contains the class of strongly contractive maps because $\phi(t)=(1-q)t$ with $0<q<1$ gives us

$$\|Tx-Ty\|\le q\|x-y\|. \tag{1.4}$$

We study the following algorithm of stochastic approximation:

$$x_{n+1}=Pr_G\left(x_n-\alpha_n S_n x_n\right), n=1,2,\cdots, x_1\in G, \tag{1.5}$$

where Pr_G is the metric projection operator from H onto G and deterministic step-parameters α_n satisfy the standard conditions:

$$\sum_{n=1}^{\infty}\alpha_n=\infty \text{ and } \sum_{n=1}^{\infty}\alpha_n^2<\infty. \tag{1.6}$$

The factor $S_n x_n$ in (1.5) is an infinite-dimensional vector of random observations of the clearance operator $F=I-T$ at random points $x_n\in G$ given for all $n\ge 1$ on the same probability space (Ω, A, P). We set

$$S_n x_n=\zeta_n+\xi_n, \tag{1.7}$$

where $\zeta_n=x_n-Tx_n$ and ξ_n is a sequence of independent random vectors with the conditions

$$E[\xi_n]=0 \text{ and } E\left[\|\xi_n\|^2\right]\le C_1, 0<C_1<\infty. \tag{1.8}$$

Here E is a symbol of the mathematical expectation. In order to calculate conditional mathematical expectations of different random variables we define the σ - subalgebra $A_n:=\sigma\left(x_1,x_2,\cdots,x_n\right)$ on (Ω, A, P). And then $E[\xi_n/A_n]$ means A_n-measurable function with the following property: for any $B\in A_n$

$$\int_B \xi_n \mathrm{d}P(\omega)=\int_B E[\xi_n/A_n]\mathrm{d}P(\omega).$$

We also assume in the sequel that ζ_n is An-measurable for all $n\ge 1$.

Let us recall the mean square convergence and almost sure (a.s.) convergence.

We say that the sequence $\{\xi_n\}$ of random variables $\xi_n(\omega)$ converges in mean square to ξ if ξ exists and

$$\lim_{n\to\infty} E\left[\|\xi_n - \xi\|^2\right] = 0.$$

The sequence ξ_n converges to ξ almost surely or with probability 1 if

$$P\left(\lim_{n\to\infty} \xi_n(\omega) = \xi(\omega)\right) = 1.$$

Almost sure convergence and convergence in mean square imply convergence in the sense of probability: The sequence $\{\xi_n\}$ of random variables $\xi_n(\omega)$ converges in the sense of probability to $\xi(\omega)$ if for all $\varepsilon > 0$

$$\lim_{n\to\infty} P\left(\|\xi_n(\omega) - \xi(\omega)\| \ge \varepsilon\right) = 0.$$

So, we consider iterative processes of stochastic approximation in the form (1.5) for finding fixed points of weakly contractive (Definition 1.1) and nonexpansive (Definition 1.3) mappings in Hilbert spaces under the conditions (1.8). We prove mean square convergence and convergence almost sure of iterative approximations and establish both asymptotic and nonasymptotic estimates of the convergence rate. Perhaps, we present here the first results of this sort for fixed point problems. Formerly the stochastic approximation methods were studied mainly to find minimal and maximal points in optimization problems (see, for example, [1-6] and references within).

2. AUXILIARY RECURRENT INEQUALITIES

Lemma 2.1. [3,4] Let $\{\mu_k\}, \{\rho_k\}$ and $\{\gamma_k\}$ be sequences of nonnegative real numbers satisfying the recurrent inequality.

$$\mu_{k+1} \le (1+\rho_k)\mu_k + \gamma_k, \; k = 1,2,\cdots \tag{2.1}$$

Assume that $\sum_{k=1}^{\infty} \rho_k < \infty$ and $\sum_{k=1}^{\infty} \gamma_k < \infty.$ Then $\{\mu_k\}$ is bounded and converges to some limit.

Lemma 2.2. [3,4] Let $\{\mu_k\}, \{\zeta_k\}, \{\alpha_k\}$ and $\{\gamma_k\}$ be sequences of nonnegative real numbers satisfying the recurrent inequality.

$$\mu_{k+1} \le (1+\rho_k)\mu_k - \alpha_k\psi(\mu_k) + \gamma_k, k = 1,2,\cdots, \tag{2.2}$$

where $\sum_{k=1}^{\infty}\alpha_k = \infty, \sum_{k=1}^{\infty}\rho_k < \infty$ and either $\sum_{k=1}^{\infty}\gamma_k < \infty$ or

$$\lim_{k\to\infty}\frac{\gamma_k}{\alpha_k} = 0. \tag{2.3}$$

Assume that $\psi(t)$ is continuous and increasing function defined on IR^+ such that ψ is positive on $IR^+/\{0\}$, $\psi(0)=0$. Then $\lim_{k\to\infty}\mu_k = 0$. There exists an infinite subsequence $\{k_l\}, l = 1,2,\cdots,$ such that

$$\mu_{k_l} \le C_0\psi^{-1}\left(\frac{1}{\sum_{k=1}^{k_l}\alpha_k} + \frac{\gamma_{k_l}}{\alpha_{k_l}}\right),$$

where $C_0 \ge \prod_{k=1}^{\infty}(1+\rho_k).$

In the following two lemmas we want to present nonasymptotic estimates for the whole sequence $\mu_k, k \ge 1$. For this the stronger requirements are made of parameters α_k and function $\psi(t)$ in the recurrent inequality.

Suppose that $\alpha(t)$ such that $\alpha(k) = \alpha_k$, $F(t)$ and $\Phi(t)$ are antiderivatives from $\alpha(t)$ and $\frac{1}{\psi(t)}$, respectively, with arbitrary constants C (without loss of generality, one can put $C = 0$), i.e.

$$F(t) = \int\alpha(t)\mathrm{d}t, \Phi(t) = \int\frac{\mathrm{d}t}{\psi(t)}.$$

Observe that $F(t)$ has the following properties:

i) $F'(t)=\alpha(t)$;

ii) $F(t)$ is strictly increasing on $[1,\infty)$ and $F(t)\to\infty$ as $t\to\infty$;

iii) The function $g(t)=\dfrac{\alpha(t)}{F(t)}=\dfrac{F'(t)}{F(t)}$ is decreasing;

iv) $G(t)=\ln F(t)\to\infty$ as $t\to\infty$.

Introduce the following denotations:

1) $\psi^{-1}(z)$ and $\Phi^{-1}(z)$ are the inverse functions to $\psi(t)$ and $\Phi(t)$, respectively;

2) $v(s)=\psi^{-1}\left[\dfrac{c_0\gamma(s-1)}{\alpha(s-1)}\right]+\gamma(s-1), s\geq 2, c_0>1$ is a fixed control parameter;

3) $u(s,C)=\Phi^{-1}\left[\Phi(C)-a\left(F(s)-F(2)\right)\right]$,

$C\geq 0, a=\dfrac{c_0-1}{c_0}<1$;

4) $w(s,v(r))=\Phi^{-1}\left[\Phi(v(r))-a\left(F(s)-F(2)\right)\right]$, $2\leq r<\infty$;

5) $Q(\bar{c})=\Phi^{-1}\left[\Phi(\bar{c})-a\left(F(2)-F(1)\right)\right]$, where

$\bar{c}>0$ is an arbitrary fixed number;

We present now the based condition (P): The graphs of the scalar functions $v(s)$ and $w(s,v(r))$ with any fixed $r\in[2,\infty)$ are intersected on the interval $[2,\infty)$ not more than at two points s_1 and s_2 (we do not consider contact points as intersection ones excepting $s=2$ if any).

For example, the graphs of the functions $v(s)$ and $w(s,v(r))$ calculated for $\alpha(s)=\dfrac{b}{s^{\kappa}}, b>0, 0<\kappa\leq 1,$ and $\psi(t)=t^{\nu},\nu\geq 1,$ satisfy the condition (P).

Lemma 2.3. [3,4] Assume that 1) the property (P) is carried out for the function $w(s,v(2)) = u(s,v(2))$ and $v(s)$; 2) $u(s,v(2)) \geq v(s)$ as $s \to \infty$; 3) the control parameter c_0 is chosen such that

$$u(s,v(2)) \geq v(s) \text{ as } s \to 2. \tag{2.4}$$

Then for the sequence $\{\mu_k\}$ generated by the inequality

$$\mu_{k+1} \leq \mu_k - \alpha_k \psi(\mu_k) + \gamma_k, k = 1,2,\cdots, \tag{2.5}$$

it follows: $\lim_{k\to\infty} \mu_k = 0$ and for all $k \geq 1$

$$\mu_k \leq u(k,C), C = \max\{Q(\mu_1), v(2)\}. \tag{2.6}$$

Lemma 2.4. [3,4] Assume that 1) the property (P) is carried out for all the function $w(s,v(2r))$ and $v(s)$; 2) $u(s,v(2)) \leq v(s)$ as $s \to \infty$. Then for the sequence $\{\mu_k\}$ generated by the inequality (2.5) $\lim_{k\to\infty} \mu_k = 0$. In additiona) if $Q(\mu_1) \leq v(2)$ and the control parameter c_0 is chosen such that $u(s,v(2)) \leq v(s)$ as $s \to 2$, then for all $k \geq 2$

$$\mu_k \leq v(k); \tag{2.7}$$

b) in all remaining cases

$$\begin{aligned} &\mu_k \leq u(k,C), C = \max\{Q(\mu_1), v(2)\}, \\ &1 \leq k \leq \bar{s}, \end{aligned} \tag{2.8}$$

$$\mu_k \leq v(k), k > \bar{s}, \tag{2.9}$$

where $\bar{s}$ is a unique root of the equation

$$u(s,C) = v(s) \tag{2.10}$$

on the interval $[2,\infty)$.

The following lemmas deal with another sort of recurrent inequalities:

Lemma 2.5. [7,8] Let $\{\mu_k\},\{\alpha_k\},\{\beta_k\}$ and $\{\gamma_k\}$ be sequences of non-negative real numbers satisfying the recurrence inequality.

$$\mu_{k+1} \le \mu_k - \alpha_k \beta_k + \gamma_k, k = 1, 2, \cdots. \tag{2.11}$$

Assume that

$$\sum_{k=1}^{\infty} \alpha_k = \infty \text{ and } \sum_{k=1}^{\infty} \gamma_k < \infty.$$

Then:

i) There exists an infinite subsequence $\{\beta_{\ell_k}\} \subset \{\beta_k\}$ such that

$$\beta_{\ell_k} \le \frac{1}{\sum_{j=1}^{\ell_k} \alpha_j}, \tag{2.12}$$

and, consequently, $\lim_{k\to\infty} \beta_{\ell_k} = 0;$

ii) if $\lim_{k\to\infty} \alpha_k = 0$ and there exists $\kappa > 0$ such that

$$|\beta_{k+1} - \beta_k| \le \kappa \alpha_k \tag{2.13}$$

for all $k \ge 1$, then $\lim_{k\to\infty} \beta_k = 0.$

Lemma 2.6. [7,8] Let $\{\mu_k\},\{\alpha_k\},\{\beta_k\}$ and $\{\gamma_k\}$ be sequences of non-negative real numbers satisfying the recurrence inequality (2.11). Assume that

$$\sum_{k=1}^{\infty} \alpha_k = \infty$$

and (2.3) is satisfied. Then there exists an infinite subsequence $\{\beta_{\ell_k}\} \subset \{\beta_k\}$ such that $\lim_{k\to\infty} \beta_{\ell_k} = 0.$

3. MEAN SQUARE CONVERGENCE OF STOCHASTIC APPROXIMATIONS

Theorem 3.1. Assume that $T: G \to H$ is a weakly contractive mapping of the class $C_{\phi(\tau)}$, $\psi_1(\tau) = \tau\phi(\tau)$ is a convex function with respect to $t = \tau^2$ and $E\left[\left\|x_1 - x^*\right\|^2\right] < \infty.$ Then the sequence $\{x_n\}$ generated by (1.5)-(1.7) converges in mean square to a unique fixed point x^* of T. There exists an infinite subsequence $\{n_l\}, l = 1, 2, \cdots,$ such that

$$E\left[\left\|x_{n_l} - x^*\right\|^2\right] \le C_0\psi^{-1}\left(\frac{1}{2\sum_{n=1}^{n_l}\alpha_n} + C_1\alpha_{n_l}\right), \tag{3.1}$$

where $\psi(\tau^2) = \psi_1(\tau)$ and some positive constant C_0 satisfies the inequality

$$\prod_{n=1}^{\infty}\left(1 + 8\alpha_n^2\right) \le C_0. \tag{3.2}$$

Remark 3.2. $1 < C_0 < \infty$ exists in view of the second condition in (1.6).

Proof. First of all, we note that the method (1.5)-(1.7) guarantees inclusion $x_n \in G$ for all $n \ge 1$. Since the metric projection operator Pr_G is nonexpansive in a Hilbert space and $x^* \in G$ exists, we can write

$$\begin{aligned}
&\left\|x_{n+1} - x^*\right\|^2 \\
&= \left\|Pr_G\left(x_n - \alpha_n\left(x_n - Tx_n + \xi_n\right)\right) - Pr_G x^*\right\|^2 \\
&\le \left\|x_n - x^*\right\|^2 - 2\alpha_n\left(\zeta_n + \xi_n, x_{n+1} - x^*\right) \\
&= \left\|x_n - x^*\right\|^2 - 2\alpha_n\left(\zeta_n, x_n - x^*\right) \\
&-2\alpha_n\left(\xi_n, x_n - x^*\right) + \left\|x_{n+1} - x_n\right\|^2.
\end{aligned} \tag{3.3}$$

Let us evaluate the first scalar product in (3.3). We have

$$\begin{aligned}&\left(x_n - Tx_n, x_n - x^*\right)\\&= \left(x_n - x^* - Tx_n + Tx^*, x_n - x^*\right)\\&= \left\|x_n - x^*\right\|^2 - \left(Tx_n - Tx^*, x_n - x^*\right)\\&\geq \left\|x_n - x^*\right\|^2 - \left\|Tx_n - Tx^*\right\|\left\|x_n - x^*\right\|\\&\geq \left\|x_n - x^*\right\|^2 - \left(\left\|x_n - x^*\right\| - \phi\left(\left\|x_n - x^*\right\|\right)\right)\left\|x_n - x^*\right\|\\&\geq \phi\left(\left\|x_n - x^*\right\|\right)\left\|x_n - x^*\right\| = \psi_1\left(\left\|x_n - x^*\right\|\right)\end{aligned} \tag{3.4}$$

We remember that $\psi\left(\tau^2\right) = \psi_1(\tau)$. Then the inequalities (3.3) and (3.4) yield

$$\begin{aligned}&\left\|x_{n+1} - x^*\right\|^2 \leq \left\|x_n - x^*\right\|^2 - 2\alpha_n \psi\left(\left\|x_n - x^*\right\|^2\right)\\&-2\alpha_n\left(\xi_n, x_n - x^*\right) + \alpha_n^2\left\|\zeta_n + \xi_n\right\|^2.\end{aligned} \tag{3.5}$$

Applying the conditional expectation with respect to A_n to the both sides of (3.5) we obtain

$$\begin{aligned}&E\left[\left\|x_{n+1} - x^*\right\|^2 / A_n\right]\\&\leq \left\|x_n - x^*\right\|^2 - 2\alpha_n \psi\left(\left\|x_n - x^*\right\|^2\right)\\&-2\alpha_n E\left[\left(\xi_n, x_n - x^*\right)/A_n\right] + \alpha_n^2 E\left[\left\|\zeta_n + \xi_n\right\|^2 / A_n\right].\end{aligned} \tag{3.6}$$

It is easy to see that

$$\begin{aligned}&E\left[\left\|\zeta_n + \xi_n^2\right\| / A_n\right]\\&\leq 2E\left[\left\|\zeta_n\right\|^2 / A_n\right] + 2E\left[\left\|\xi_n\right\|^2 / A_n\right]\\&= 2\left\|\zeta_n\right\|^2 + 2E\left[\left\|\xi_n\right\|^2 / A_n\right].\end{aligned} \tag{3.7}$$

Since T is weakly contractive and therefore nonexpansive, one gets

$$\left\|x_n - Tx_n\right\|^2 \le 4\left\|x_n - x^*\right\|^2.$$

Taking into account (3.7), the inequality (3.9) is estimated as follows:

$$E\left[\left\|x_{n+1} - x^*\right\|^2 / A_n\right]$$
$$\le \left(1 + 8\alpha_n^2\right)\left\|x_n - x^*\right\|^2 - 2\alpha_n\psi\left(\left\|x_n - x^*\right\|^2\right)$$
$$-2\alpha_n E\left[\left(\xi_n, x_n - x^*\right)/A_n\right] + 2\alpha_n^2 E\left[\left\|\xi_n\right\|^2 / A_n\right]. \tag{3.8}$$

Now the unconditional expectation implies

$$E\left[\left\|x_{n+1} - x^*\right\|^2\right]$$
$$\le \left(1 + 8\alpha_n^2\right)E\left[\left\|x_n - x^*\right\|^2\right] - 2\alpha_n E\left[\psi\left(\left\|x_n - x^*\right\|^2\right)\right]$$
$$-2\alpha_n E\left[\left(\xi_n, x_n - x^*\right)\right] + 2\alpha_n^2 E\left[\left\|\xi_n\right\|^2\right] \tag{3.9}$$

Next we need the Jensen inequality for a convex function $\psi\left(\left\|x_n - x^*\right\|^2\right)$:

$$E\left[\psi\left(\left\|x_n - x^*\right\|^2\right)\right] \le \psi\left(E\left[\left\|x_n - x^*\right\|^2\right]\right)$$

(see [9,10]). This allows us to rewrite (3.9) in the form

$$E\left[\left\|x_{n+1} - x^*\right\|^2\right]$$
$$\le \left(1 + 8\alpha_n^2\right)E\left[\left\|x_n - x^*\right\|^2\right]$$
$$-2\alpha_n\psi\left(E\left[\left\|x_n - x^*\right\|^2\right]\right) + 2\alpha_n^2 E\left[\left\|\xi_n\right\|^2\right] \tag{3.10}$$

because of

$$E\left[\left(\xi_n, x_n - x^*\right)\right] = 0.$$

Denoting $\lambda_n = E\left[\left\|x_n - x^*\right\|^2\right]$ we have

$$\lambda_{n+1} \le \left(1+8\alpha_n^2\right)\lambda_n - 2\alpha_n\psi\left(\lambda_n\right) + 2C_1\alpha_n^2, \tag{3.11}$$

where in view of Definition 1.1, $\psi(\lambda)$ is a continuous and increasing function with $\psi(0)=0.$ Due to (6), from Lemma 2.2 it follows

$$\lim_{n\to\infty} E\left[\left\|x_n - x^*\right\|^2\right] = 0$$

and the estimate (3.1) holds too. The theorem is proved. □

Remark 3.3. If a fixed point of weakly contractive mapping $T:G\to H$ exists, then it is unique [11].

Remark 3.4. The following example was presented in [11]: Let $Tx=\sin x,$ $G=[0,1]$ and $T:G\to G.$ It has been shown in [11] that

$$\left|\sin x - \sin y\right| \le \left|x-y\right| - \frac{1}{8}\left|x-y\right|^3$$

for all $0\le x\le y\le 1.$ Then

$$\phi(\tau)=\frac{1}{8}\tau^3, \psi_1(\tau)=\frac{1}{8}\tau^4 \text{ and } \psi(t)=\frac{1}{8}t^2.$$

Definition 3.5. Let a nonexpansive mapping

$T:G\to H$ have a unique fixed point x^*. T is said to be weakly sub-contractive on the closed convex subset $G\subset H$ if there exists continuous and increasing function $\psi(t)$ defined on IR^+ such that ψ is positive on $IR^+/\{0\}$, $\psi(0)=0$, $\lim_{t\to+\infty}\phi(t)=+\infty,$ such that for all

$$x\in G.$$

$$\left\langle x - Tx, x - x^*\right\rangle \ge \psi\left(\left\|x-x^*\right\|^2\right). \tag{3.12}$$

Theorem 3.6. Assume that a mapping $T:G \to H$ is weakly sub-contractive and the function $\psi(t)$ in (3.12) is convex on IR^+ Then the results of Theorem 3.1 holds for the sequence $\{x_n\}$ generated by (1.5)-(1.7).

The second inequality in (1.6) can be omitted if we assume not less than linear growth of $\psi(\lambda)$ "on infinity" and put $\alpha_n \to 0$ as $n \to \infty$

Theorem 3.7. Assume that a mapping $T:G \to H$ is weakly sub-contractive and the function $\psi(t)$ in (3.12) is convex on IR^+ Suppose that instead of (1.6) the conditions

$$\lim_{n\to\infty} \alpha_n = 0 \text{ and } \sum_{n=1}^{\infty} \alpha_n = \infty. \tag{3.13}$$

hold. In addition, let $\alpha_n \le \bar{\alpha} < 0.5$ and

$$\lim_{\lambda\to\infty} \frac{\psi(\lambda)}{\lambda} \ge 4\bar{\alpha}. \tag{3.14}$$

Then the sequence $\{x_n\}$ generated by (1.5), (1.7) and (3.13) converges in mean square to x^*. There exists an infinite subsequence $\{n_l\}, l = 1, 2, \cdots,$ such that

$$E\left[\left\|x_{n_l} - x^*\right\|^2\right] \le \psi^{-1}\left(\frac{1}{2\sum_{n=1}^{n_l} \alpha_n} + \frac{C_3 \alpha_{n_l}}{2}\right).$$

where

$$C_3 = 2C_1 + 8\max\left\{\left\|x^* - x_1\right\|^2, C_2\right\},$$
$$C_2 = \lambda_* + 2C_1\bar{\alpha}^2 + 8\bar{\alpha}^2\lambda_*,$$
$$\lambda_* = \psi^{-1}\left(C_1\bar{\alpha} + 4\bar{\alpha}\lambda_*\right). \tag{3.15}$$

Proof. Consider the inequality (11) in the form

$$\lambda_{n+1} \le \lambda_n - 2\alpha_n\psi\left(\lambda_n\right) + 2C_1\alpha_n^2 + 8\alpha_n^2\lambda_n, \tag{3.16}$$

where $\lambda_n = E\left[\left\|x_n - x^*\right\|^2\right].$ Observe that it is derived by making use of (3.4) and the nonexpansivity property of $T.$ We shall show that λ_n are bounded for all $n \in I = \{1, 2, \cdots\}.$ Indeed, since $\psi(\lambda)$ is a convex increasing continuous function, we conclude that $\psi_1(\lambda) = \frac{\psi(\lambda)}{\lambda}$ is nondecreasing and since (3.14) holdsthe inequality $\psi(\lambda) \le C_1\bar{\alpha} + 4\bar{\alpha}\lambda$ has a solution $\lambda \le \lambda_*,$ where λ_* is the unique root of the scalar equ0 ation $\psi(\lambda) = C_1\bar{\alpha} + 4\bar{\alpha}\lambda.$ Together with this, (3.4) and (3.14) are co-ordinated by the parameter $\bar{\alpha} < 0.5.$

Only one alternative can happen for each $n \in I:$ either

$$H_1 := -2\alpha_n\psi(\lambda_n) + 2C_1\alpha_n^2 + 8\alpha_n^2\lambda_n \ge 0$$

or

$$H_2 := -2\alpha_n\psi(\lambda_n) + 2C_1\alpha_n^2 + 8\alpha_n^2\lambda_n < 0.$$

Denote $I_1 = \{n \in I | H_1 \text{ is true}\}$ and $I_2 = \{n \in I | H_2 \text{ is true}\}$. It is clear that $I_1 \cup I_2 = I.$ From the hypothesis $H_1,$ it arises

$$\psi(\lambda_n) \le C_1\bar{\alpha} + 4\bar{\alpha}\lambda_n,$$

and then $\lambda_n \le \lambda_*$ for all $n \in I_1.$ From the hypothesis $H_2,$ we have: $\lambda_{n+1} < \lambda_n$ for all $n \in I_2.$ Consider all the possible cases:

1) $I_2 = \varnothing.$ Then $\lambda_n \le \lambda_*$ for all $n \in I.$

2) $I_1 = \varnothing.$ Then $\lambda_n \le \lambda_1$ for all $n \in I.$

3) Let $I_1 = \{1, 2, \cdots, N_0\}$ and $I_2 = \{N_0 + 1, N_0 + 2, \cdots\}.$ Then $\lambda_n \le \lambda_*$ for $n = 1, 2, \cdots, N_0.$ By (3.16), $\lambda_{N_0+1} \le C_2.$ It is obvious that $\lambda_n \le C_2$ for $N_0 + 2, N_0 + 3, \cdots.$ Therefore, $\lambda_n \le C_2$ for all $n \in I.$

4) Let $I_2 = \{1, 2, \cdots, N_0\}$ and $I_1 = \{N_0 + 1, N_0 + 2, \cdots\}.$ Then $\lambda_n \le \lambda_1$ for $n = 1, 2, \cdots, N_0.$ and $\lambda_n \le \lambda_*$ for $N_0 + 1, N_0 + 2, \cdots$ Thus, $\lambda_n \le \lambda_*$ for all $n \in I.$

5) Let I_1 and I_2 be unbounded sets. Consider an arbitrary interval

$I_s = [n_s + 1, n_{s+1} - 1] \subset I_2,$

where $n_s, n_{s+1} \in I_1, s = 1,3,5,\cdots$. It is easy to be sure that $\lambda_{n_s} \le \lambda_*$ and $\lambda_n \le C_2$ for all $n \in I_s$.

6) The other situations of bounded and unbounded sets I_1 and I_2 are covered by the items 1)-5). Consequently, we have the final result: $\lambda_n \le \max\{\lambda_1, C_2\}$ for all $n \in I$.

Thus, we obtain the inequality

$$\lambda_{n+1} \le \lambda_n - 2\alpha_n \psi(\lambda_n) + C_3 \alpha_n^2, \tag{3.17}$$

where C_3 is defined by (3.15). Now Lemma 2.2 with the condition (2.3) implies the result. □

Remark 3.8. For a linear function $\psi(\tau) = c\tau$ which is convex and concave at the same time we suppose $2 \ge c > 4\bar{\alpha}$.

Remark 3.9. If G is bounded or more generally $\{x_n\}$ is bounded, then the inequality (3.17) (with some different constant C_3) immediately follows from (3.16).

4. ESTIMATES OF THE MEAN SQUARE CONVERGENCE RATE

Using Lemmas 2.3 and 2.4 we are able to give two general theorems on the nonasymptotic estimates of the mean square convergence rate for sequence $\{x_n\}$ generated by the stochastic approximation algorithm (1.5)- (1.7).

Again we introduce denotations 1)-5) from Section 2 induced now by the recurrent inequality (3.11):

1) $\psi^{-1}(z)$ and $\Phi^{-1}(z)$ are the inverse functions to $\psi(t)$ and $\Phi(t)$, respectively;

2) $v(s) = \psi^{-1}[C_1 c_0 \alpha(s-1)] + 2C_1 \alpha^2(s-1), c_0 > 1$ is a fixed control parameter;

3) $u(s, C_2) = \Phi^{-1}[\Phi(C_2) - a(F(s) - F(2))], C_2 > 0,$

$$a = \frac{c_0 - 1}{c_0} < 1$$

4) $w(s, v(r)) = \Phi^{-1}\left[\Phi(v(r)) - a(F(s) - F(2))\right],$

$2 \le s, r < \infty;$

5) $Q = \Phi^{-1}\left[\Phi\left(\|x_1 - x^*\|^2\right) - a(F(2) - F(1))\right].$

Introduce also the basic condition (P).

Theorem 4.1. Assume that all the conditions of Theorem 3.1 are fulfiled and

i) the condition (P) holds for the functions

$w(s, v(2)) = u(s, v(2))$ and $v(s)$;

ii) $u(s, v(2)) \ge v(s)$ as $s \to \infty$;

iii) $c_0 > 1$ is chosen such that $u(s, v(2)) \ge v(s)$ as $s \to 2$.

Then the sequence $\{x_n\}$ generated by (1.5)-(1.7) converges in average to a unique fixed point x^* of T and for all $n \ge 1$

$$E\left[\|x_n - x^*\|^2\right] \le C_0 u(n, C_2),$$
$$C_2 = \max\{Q, v(2)\}. \tag{4.1}$$

Theorem 4.2. Assume that all the conditions of Theorem 3.1 are fulfiled and i) the condition (P) holds for the functions $v(s)$

and $w(s, v(z))$ with any fixed $z \in [2, \infty)$;

ii) $u(s, v(2)) \le v(s)$ as $s \to \infty$;

iii) If $Q \le v(2)$ and $c_0 > 1$ is chosen such that $u(s, v(2)) \le v(s)$ as $s \to 2$, then the sequence $\{x_n\}$ generated by (1.5)-(1.7) converges in average to a unique fixed point x^* of T and for all $n \ge 1$

$$E\left[\|x_n - x^*\|^2\right] \le C_0 v(n); \tag{4.2}$$

iv) In all the remaining cases, (4.1) holds for $1 \le n \le \bar{s}$ and (4.2) for $n \ge \bar{s}$, where $\bar{s}$ is a unique root of the equation $u(s, C_2) = v(s)$ on the interval $[2, \infty)$.

Let us provide the examples of functions $\psi(\tau)$ and $\Phi(\lambda)$ suitable for Theorems 4.1 and 4.2 (see [12,13]).

1) Below in Corollaries 4.3-4.6 we use the functions $\psi(\tau) = \tau^{\nu}$ with $\nu \ge 1$. For them

$$\Phi(\lambda) = \begin{cases} \ln \lambda, \text{ if } \nu = 1, \\ \dfrac{\lambda^{1-\nu}}{1-\nu}, \text{ if } \nu \ne 1, \end{cases} \tag{4.3}$$

and

$$\Phi^{-1}(z) = \begin{cases} \exp z, \text{ if } \nu = 1, \\ \left[(1-\nu) z\right]^{\frac{1}{1-\nu}}, \text{ if } \nu \ne 1, \end{cases} \tag{4.4}$$

2) If $\psi(\tau) = \exp \tau - 1, \tau > 0,$ then

$$\Phi(\tau) = \ln(1 - \exp \tau) \text{ and } \Phi^{-1}(z) = -\ln(1 - \exp z).$$

3) If $\psi(\tau) = \dfrac{\tau}{\tau + 1}, \tau > 0,$ then

$$\Phi(\tau) = \ln \tau + \tau \text{ and } \Phi^{-1}(z) \le \ln(1 + \exp z).$$

4) If $\psi(\tau) = \dfrac{\tau^2}{\tau + 1}, \tau > 0,$ then

$$\Phi(\tau) = \ln \tau - \frac{1}{\tau}.$$

In this example we are unable to define $\Phi^{-1}(z)$ in analitical form, therefore suggest to calculate it numerically by computer.

We next present very important corollaries from Theorems 4.1 and 4.2, where their assumptions automatically guarantee accomplishment of the condition (P) (see [4]). The functions $\psi(\tau)$ coincide with the point 1) above.

Corollary 4.3. Assume that $T:G\to H$ is a strongly contractive mapping, that is, (1.4) is satisfied with

$0<q<1$. Let in (1.5) $\alpha_n=\frac{b}{n}, b>0.$ Then

$$F(t)=b\ln t, \Phi(\tau)=\frac{\ln\tau}{q_1}, \Phi^{-1}(z)=e^{q_1 z}, q_1=1-q,$$

$$v(s)=\left(\frac{2c_0C_1}{b}+2C_1\right)\frac{1}{s-1}, u(s,C)=C\left(\frac{2}{s}\right)^{abq_1},$$

$$Q=\left\|x_1-x^*\right\|^2\left(\frac{1}{2}\right)^{abq_1}.$$

I. Suppose that $b>\frac{1}{q_1}$ and $c_0\geq\frac{bq_1}{bq_1-1}$. Then

$\lim_{n\to\infty}E\left[\left\|x_n-x^*\right\|^2\right]=0$ and 1) If $Q\leq v(2)$ and $c_0\geq\frac{bq_1}{bq_1-2}$, we have for all $n>1$

$$E\left[\left\|x_n-x^*\right\|^2\right]\leq v(n); \tag{4.5}$$

2) In all the remain cases

$$E\left[\left\|x_n-x^*\right\|^2\right]\leq u(n,C),$$
$$C=\max\left\{Q,v(2)\right\}, 1<n\leq\bar{s}, \tag{4.6}$$

and

$$E\left[\left\|x_n-x^*\right\|^2\right]\leq v(s), n\geq\bar{s}, \tag{4.7}$$

where $\bar{s}$ is a unique root of the equation

$u(s,C)=v(s)$ on the interval $[2,\infty)$.

II. Suppose that $\frac{1}{2q_1}<b<\frac{1}{q_1}$ and $c_0\geq\frac{bq_1}{1-bq_1}>1.$

Then $\lim_{n\to\infty}E\left[\left\|x_n-x^*\right\|^2\right]=0$ and the estimate (4.6) holds for all $n\geq 1$.

Corollary 4.4. Assume that $T:G\to H$ is a strongly contractive mapping, that is, in (1.4) is satisfied with

$0<q<1$. Let in (1.5) $\alpha_n=\frac{b}{n^{\kappa}},b>0,0<\kappa<1.$ Then

$$F(t)=\frac{b}{1-\kappa}t^{1-\kappa},\Phi(\tau)=\frac{\ln\tau}{q_1},$$

$$\Phi^{-1}(z)=\left(\frac{1-\kappa}{b}z\right)^{\frac{1}{1-\kappa}},q_1=1-q,$$

$$v(s)=\left(\frac{2c_0C_1}{b}+2C_1\right)\frac{1}{s-1},$$

$$u(s,C)=C\exp\left\{-\frac{abq_1}{1-\kappa}\left(s^{1-\kappa}-2^{1-\kappa}\right)\right\},$$

$$Q=\left\|x_1-x^*\right\|^2\exp\left\{-\frac{abq_1}{1-\kappa}\left(2^{1-\kappa}-1\right)\right\}.$$

Suppose that $c_0\geq\frac{bq}{bq-1}$. Then $\lim_{n\to\infty}E\left[\left\|x_n-x^*\right\|^2\right]=0$

and 1) If $Q\leq v(2)$ and $c_0\geq\frac{bq_1}{bq_1-2}$, we have for all

$n>1$

$$E\left[\left\|x_n-x^*\right\|^2\right]\leq v(n); \tag{4.8}$$

2) In all the remain cases the estimates (4.6) and (4.7) hold.

Corollary 4.5. Assume that $T:G\to H$ is a weakly contractive mapping of the class $C_{\phi(t)}=t^{2\nu-1},\nu>1,$ that is, in Theorem 3.1 $\psi(\tau)=\tau^{\nu}$. Let in (1.5)

$\alpha_n=\frac{b}{n},b>0.$ Then

$$F(t)=b\ln t,\Phi(\tau)=\frac{\tau^{1-\nu}}{1-\nu},\Phi^{-1}(z)=\left[(1-\nu)z\right]^{\frac{1}{1-\nu}},$$

$$u(s,C)=C\left[1+(\nu-1)C^{\nu-1}ab\ln\frac{s}{2}\right]^{-\frac{1}{\nu-1}},$$

$$Q=\left\|x_1-x^*\right\|^2\left[1+(\nu-1)\left\|x_1-x^*\right\|^{2(\nu-1)}ab\ln 2\right]^{-\frac{1}{\nu-1}}.$$

If $c_0 > 1$ is chosen from the condition

$$\left(\frac{2c_0C_1}{b}+2C_1\right)^{\nu-1} ab \le \nu^{-1},$$

then $\lim_{n\to\infty} E\left[\left\|x_n - x^*\right\|^2\right]=0$ and for all $n>1$

$$E\left[\left\|x_n - x^*\right\|^2\right] \le u(n,C),$$

$$C=\max\left\{Q,\left(\frac{2C_1c_0}{b}\right)^{\frac{1}{\nu}}+2C_1\right\}.$$

(4.9)

Corollary 4.6. Assume that $T:G\to H$ is a weakly contractive mapping of the class $C_{\phi(t)}=t^{2\nu-1},\nu>1,$ that is, in Theorem 3.1 $\psi(\tau)=\tau^{\nu}$. Let in (1.5)

$\alpha_n=\frac{b}{n^{\kappa}}, b>0, 0<\kappa<1.$ Then

$$F(t)=\frac{b}{1-\kappa}t^{1-\kappa},\Phi(\tau)=\frac{\tau^{1-\nu}}{1-\nu},\Phi^{-1}(z)=\left[(1-\nu)z\right]^{\frac{1}{1-\nu}},$$

$$v(s)=\left[\left(\frac{2c_0C_1}{b}\right)^{\frac{1}{\nu}}+2C_1\right]\left(\frac{1}{s-1}\right)^{\frac{\kappa}{\nu}},$$

$$u(s,C)=C\left[1+\frac{\nu-1}{1-\kappa}C^{\nu-1}ab\left(s^{1-\kappa}-2^{1-\kappa}\right)\right]^{-\frac{1}{\nu-1}},$$

$$Q=\left\|x_1-x^*\right\|^2\left[1+\frac{\nu-1}{1-\kappa}\left\|x_1-x^*\right\|^{2(\nu-1)}ab\left(2^{1-\kappa}-1\right)\right]^{-\frac{1}{\nu-1}}.$$

I. Suppose that

$$\frac{\kappa}{\nu}<\frac{1-\kappa}{\nu-1}.$$

1) If $Q\le v(2)$ and $c_0>1$ is chosen from the condition

$$\left[\left(\frac{2c_0C_1}{b}\right)^{\frac{1}{\nu}}+2C_1\right]^{\nu-1} ab \ge \frac{2^{\kappa}\kappa}{\nu},$$

then $\lim_{n\to\infty} E\left[\left\|x_n - x^*\right\|^2\right] = 0$ and for all $n > 1$

$$E\left[\left\|x_n - x^*\right\|^2\right] \le v(n). \tag{4.10}$$

2) In all the remain cases the estimates (4.6) and (4.7) hold.

II. Suppose that

$$\frac{\kappa}{\nu} \ge \frac{1-\kappa}{\nu-1}.$$

If $c_0 > 1$ is chosen from the condition

$$\left[\left(\frac{2c_0C_1}{b}\right)^{\frac{1}{\nu}} + 2C_1\right]^{\nu-1} ab \le \frac{\kappa}{\nu},$$

then $\lim_{n\to\infty} E\left[\left\|x_n - x^*\right\|^2\right] = 0$ and for all $n \ge 1$

$$E\left[\left\|x_n - x^*\right\|^2\right] \le u(n,C), C = \max\{Q, v(2)\}. \tag{4.11}$$

In addition to the examples presented in this section, we produce the functions $\psi(\tau)$ and $\Phi(\lambda)$ which have $\tau = 0$ as a tangency point of the infinite degree multiplicity and given logarithmic estimates of the convergence rate.

We define the function $\psi(\tau)$ by the following way:

$$\psi(\tau) = -\exp(-f(\tau))\left[f'(\tau)\right]^{-1}, \tau > 0,$$

where $f(\tau)$ is differentiable and decreasing function,

$\lim_{\tau\to 0} f(\tau) = \infty,$ and

$$\psi^{(l)}(0) = \lim_{\tau\to 0+} \psi^{(l)}(\tau), l = 0,1,2,\cdots,$$

where (l) denote the derivative degrees of the function

$\psi(\tau), \psi^{(0)}(\tau) = \psi(\tau).$ It is easy to see that

$\Phi(\tau) = -\exp f(\tau)$

and

$\Phi^{-1}(z)=f^{-1}(\ln(-z)).$

In particulari) $f(\tau)=\frac{1}{\tau},\psi(\tau)=\tau^2\exp\left(-\frac{1}{\tau}\right).$ We have

$\Phi(\tau)=-\exp\frac{1}{\tau}$ and $\Phi^{-1}(z)=\frac{1}{\ln(-z)}.$ We have to verify that $\psi(\tau)$ is convex. In fact, it is true because

$$\frac{d^2\psi(\tau)}{d\tau}=\left(2+\frac{2\tau+1}{\tau^2}\right)\exp\left(-\frac{1}{\tau}\right)>0,\forall\tau>0.$$

Beside this, it is easy to see that $\psi(\tau)\leq\tau,$ at least, on the interval $[0,1]$. In the next examples we leave to readers to check these properties.

ii) $f(\tau)=\frac{1}{\tau^s},\ s\geq 1,\ \psi(\tau)=\exp\left(-\frac{1}{\tau^s}\right)\tau^{s+1}.$

We have $\Phi(\tau)=-\exp\frac{1}{\tau^s}$ and $\Phi^{-1}(z)=(\ln(-z))^{-\frac{1}{s}}.$

iii) $f(\tau)=\exp\frac{1}{\tau^s},\psi(\tau)=\exp\left(-\left[\frac{1}{\tau^s}+\exp\left(\frac{1}{\tau^s}\right)\right]\right)\tau^{s+1}$

We have

$\Phi(\tau)=-\exp\left(\exp\frac{1}{\tau^s}\right)$ and $\Phi^{-1}(z)=(\ln\ln(-z))^{-\frac{1}{s}}.$

5. ALMOST SURE CONVERGENCE OF STOCHASTIC APPROXIMATIONS FOR NONEXPANSIVE MAPPINGS

Consider next the almost surely convergence of stochastic approximations. First of all, we need the stochastic analogy of Lemma 2.5:

Lemma 5.1. Let $\{\alpha_k\}$ be sequences of non-negative real numbers and $\{\beta_k\}$ be sequence of random A_k measurable variables, a.s. nonnegative for all $k\geq 1$. Assume that

$$\sum_{k=1}^{\infty}\alpha_k=\infty \text{ and } \sum_{k=1}^{\infty}\alpha_k\beta_k<\infty.$$

If $\lim_{k\to\infty} \alpha_k = 0$ and there exists $c > 0$ such that for all

$k \geq 1$

$$\left|\beta_{k+1} - E\left[\beta_k / A_n\right]\right| \leq c\alpha_k \text{ a.s.,} \tag{5.1}$$

then $\lim_{k\to\infty} \beta_k = 0$ a.s.

The proof can be provided by the scheme of nonstochastic case (see Proposition 2 in [8]) or as it was done in [5].

We need also the following lemma from [14] as applied to our case of Hilbert spaces (the concepts of modulus of convexity $\delta_B(\varepsilon)$ of Banach spaces B or Hilbert spaces H can be found in [15] and [16]).

Lemma 5.2. If $F = I - T$ with a nonexpansive mapping $T: G \to H$, then for all $x, y \in D(T)$,

$$(Fx - Fy, x - y) \geq R_1^2 \delta_H \left(\frac{\|Fx - Fy\|}{2R_1} \right),$$

where

$$R_1 = \sqrt{2^{-1}\left(\|x-y\|^2 + \|Tx - Ty\|^2\right)} \leq \|x - y\|.$$

If $\|x\| \leq R$ and $\|y\| \leq R$ with $x, y \in D(T)$, then $R_1 \leq 2R$ and

$$(Fx - Fy, x - y) \geq L^{-1} R^2 \delta_H \left(\frac{\|Fx - Fy\|}{4R} \right).$$

Theorem 5.3. Assume that a mapping $T: G \to H$ is nonexpansive and its fixed point set N is nonempty. If (1.8) holds and $E\left[\|\xi_n\| / A_n\right] \leq C_0$, then the sequence $\{x_n\}$ generated by (1.5)-(1.7) weakly almost surely converges to some $\tilde{x} \in N$.

Proof. Let $x^* \in N$. We next use Lemma 5.2 and the estimate (see [17], p. 49)

$$\delta_H(\varepsilon) \geq \frac{\varepsilon^2}{8}$$

to get

$$\left(x-Tx,x-x^{*}\right)\geq R_{1}^{2}\delta_{H}\left(\frac{\|x-Tx\|}{2R_{1}}\right)\geq\frac{\|x-Tx\|^{2}}{32}.$$

In this case the inequality (3.3) implies

$$\begin{aligned}&\left\|x_{n+1}-x^{*}\right\|^{2}\\&\leq\left(1+8\alpha_{n}^{2}\right)\left\|x_{n}-x^{*}\right\|^{2}-(32)^{-1}\left\|\alpha_{n}x_{n}-Tx_{n}\right\|^{2}\\&-2\alpha_{n}\left(\xi_{n},x_{n}-x^{*}\right)+2\alpha_{n}^{2}\left\|\xi_{n}\right\|^{2}.\end{aligned}\tag{5.2}$$

Similarly to (3.10), we have

$$\begin{aligned}&E\left[\left\|x_{n+1}-x^{*}\right\|^{2}/A_{n}\right]\\&\leq\left(1+8\alpha_{n}^{2}\right)\left\|x_{n}-x^{*}\right\|^{2}-2\alpha_{n}E\left[\left\|x_{n}-Tx_{n}\right\|^{2}/A_{n}\right]\\&-2\alpha_{n}E\left[\left(\xi_{n},x_{n}-x^{*}\right)/A_{n}\right]+2\alpha_{n}^{2}E\left[\left\|\xi_{n}\right\|^{2}/A_{n}\right].\end{aligned}\tag{5.3}$$

Denote $\lambda_n=E\left[\left\|x_n-x^*\right\|^2\right]$ and $\beta_n=E\left[\left\|x_n-Tx_n\right\|^2\right]$

and apply the unconditional expectation to both sides of (5.3). Then

$$\lambda_{n+1}\leq\left(1+8\alpha_{n}^{2}\right)\lambda_{n}-2\alpha_{n}\beta_{n}+2C_{1}\alpha_{n}^{2}.\tag{5.4}$$

It follows from this that

$$\lambda_{n+1}\leq\left(1+8\alpha_{n}^{2}\right)\lambda_{n}+2C_{1}\alpha_{n}^{2}.$$

Since $\sum_1^\infty\alpha_n^2<\infty$ and due to Lemma 2.1, we conclude that $\{\lambda_n\}$ is bounded. Consequently, $\{x_n\}$ is bounded a.s. that follows from the theory of convergent quasimartingales (see [5,18]).

We now need Lemma 5.1. It is not difficult to see that

$$\sum_{1}^{\infty}\alpha_{n}\beta_{n}=\sum_{1}^{\infty}\alpha_{n}E\left[\left\|x_{n}-Tx_{n}\right\|^{2}\right]<\infty.\tag{5.5}$$

The last gives us

$$\sum_{1}^{\infty}\alpha_{n}\left\|x_{n}-Tx_{n}\right\|^{2}<\infty\text{ a.s.}$$

Next we evaluate the following difference:

$$\begin{aligned}&\|x_{n+1}-Tx_{n+1}\|^2-\|x_n-Tx_n\|^2\\&\le\left(\|x_{n+1}-Tx_{n+1}\|+\|x_n-Tx_n\|\right)\\&\times\left(\|x_{n+1}-Tx_{n+1}\|-\|x_n-Tx_n\|\right)\end{aligned}$$

It is easy to see that $\|x_n-Tx_n\|$ is bounded a.s. Indeed, since $\|x_n\|\le C_1$ a.s., there exists a constant $C_2>0$ such that

$$\|x_n-Tx_n\|=2\|x_n-x^*\|\le C_2 \text{ a.s.}$$

Therefore

$$\|x_{n+1}-Tx_{n+1}\|+\|x_n-Tx_n\|\le 2C_2 \text{ a.s.}$$

It is obviously that

$$\begin{aligned}&\|x_{n+1}-Tx_{n+1}\|-\|x_n-Tx_n\|\\&\le\|(x_{n+1}-x_n)+(Tx_n-Tx_{n+1})\|\le 2\|x_{n+1}-x_n\|\\&=2\|\mathrm{Pr}_G\, x_n-\alpha_n(x_n-Tx_n+\xi_n)-\mathrm{Pr}_G\, x_n\|\\&\le 2\alpha_n\|x_n-Tx_n\|+2\alpha_n\|\xi_n\|.\end{aligned}$$

Thus,

$$\begin{aligned}&E\left[\|x_{n+1}-Tx_{n+1}\|^2/A_n\right]-\|x_n-Tx_n\|^2\\&\le 4\alpha_n C_2^2+4\alpha_n C_2 E\left[\|\xi_n\|/A_n\right]\\&\le 4C_2(C_2+C_0)\alpha_n \text{ a.s..}\end{aligned}$$

By Lemma 5.1, $\|x_n-Tx_n\|\to 0$ a.s. as $n\to\infty$.

Since $\{x_n\}$ is bounded a.s., there is a subsequence $\{x_{n_k}\}$ weakly convergent to some point $\tilde{x}$. Since is convex and closed, consequently, weakly closed, we assert that $\tilde{x}\in G$. It is known that a nonexpansive mapping T is weakly demiclosed, therefore, $\tilde{x}\in N$ a.s. Weak almost surely convergence of whole sequence $\{x_n\}$ is shown by the standard way [8]. □

Corollary 5.4. Assume that $T:G\to H$ is a weakly contractive mapping of the class $C_{\phi(t)}$. If $\sum_1^\infty \alpha_n=\infty,\ \sum_1^\infty \alpha_n^2<\infty$ and $E[\|\xi_n\|/A_n]\le C_0$, then the sequence $\{x_n\}$

generated by (1.5)-(1.7) strongly almost surely converges to unique fixed point x^* of T.

Proof. We have from (3.4)

$$\|x_n - Tx_n\| \|x_n - x^*\| \geq \psi\left(\|x_n - x^*\|^2\right). \tag{5.6}$$

Since $\{x_n\}$ is bounded a.s. and $\|x_n - Tx_n\| \to 0$ a.s.

as $n \to \infty$, we conclude that $\psi\left(\|x_n - x^*\|^2\right) \to 0$ a.s.

The proof follows due to the properties of the function ψ. □

Remark 5.5. It is clear that all the results remain still valid for self-mappings $T: G \to G$. However, in this case, unlike any deterministic situation, the algorithm (1.5)-(1.7) must use the projection operator Pr_G because the vector $v_n = x_n - \alpha_n S_n x_n$ not always belongs to G. If $T_n x_n = Tx_n + \xi_n \in G$ for all $n \geq 1$ and $0 < \alpha_n < 1$, then (1.5) can be replaced by

$$x_{n+1} = (1 - \alpha_n) x_n + \alpha_n T_n x_n, n = 1, 2, \cdots, x_1 \in G.$$

REFERENCES

1. M. T. Wasan, "Stochastic Approximation," Cambridge University Press, Cambridge, 1969.

2. M. B. Nevelson and R. Z. Hasminsky, "Stochastic Approximation and Recursive Estimation," AMS Providence, Rhode Island, 1973.

3. Ya. Alber and S. Shilman, "Nonasymptotic Estimates of the Convergence Rate of Stochastic Iterative Algorithms," Automation and Remote Control, Vol. 42, 1981, pp. 32-41.

4. Ya. Alber and S. Shilman, "General Nonasymptotic Estimates of the Convergence Rate of Iterative Stochastic Algorithms," USSR Computational Mathematics and Mathematical Physics, Vol. 25, No. 2, 1985, pp. 13-20.

5. K. Barty, J.-S. Roy and C. Strugarek, "Hilbert-Valued Perturbed Subgradient Algorithms," Mathematics of Operation Research, Vol. 32, No. 3, 2007, pp. 551-562.

6. X. Chen and H. White, "Asymptotic Properties of Some Projection-Based Robbins-Monro Procedures in a Hilbert Space," Studies in Nonlinear Dynamics & Econometrics, Vol. 6, No. 1, 2002, pp. 1-53.

7. Ya. I. Alber and A. N. Iusem, "Extension of Subgradient Techniques for Nonsmooth Optimization in Banach Spaces," Set-Valued Analysis, Vol. 9, No. 4, 2001, pp. 315-335.

8. Ya. Alber, A. Iusem and M. Solodov, "On the Projection Subgradient Method for Nonsmooth Convex Optimization in a Hilbert Space," Mathematical Programming, Vol. 81, No. 1, 1998, pp. 23-35.

9. P. Dempster, N. M. Laird and D. B. Rubin, "Maximal Likelihood from Incomplete Data via the EM Algorithm," Journal of the Royal Statistical Society, Vol. 39, No. 1, 1977, pp. 185-197.

10. W. Rudin, "Real and Complex Analysis," McGraw-Hill, New York, 1978.

11. Ya. I. Alber and S. Guerre-Delabriere, "Principle of Weakly Contractive Maps in Hilbert Spaces," Operator Theory, Advances and Applications, Vol. 98, 1997, pp. 7- 22.

12. Ya. I. Alber, "Recurrence Relations and Variational Inequalities," Soviet Mathematics Doklady, Vol. 27, 1983, pp. 511-517.

13. Ya. Alber, S. Guerre-Delabriere and L. Zelenko, "The Principle of Weakly Contractive Maps in Metric Spaces," Communications on Applied Nonlinear Analysis, Vol. 5, No. 1, 1998, pp. 45-68.

14. Ya. I. Alber, "New Results in Fixed Point Theory," Preprint, Haifa Technion, 2000.

15. J. Diestel, "The Geometry of Banach Spaces," Lecture Notes in Mathematics, No. 485, Springer, Berlin, 1975.

16. T. Figiel, "On the Moduli of Convexity and Smoothness," Studia Mathematica, Vol. 56, No. 2, 1976, pp. 121-155.

17. Ya. Alber and I. Ryazantseva, "Nonlinear Ill-Posed Problems of Monotone Type," Springer, Berlin, 2006.

18. M. Métivier, "Semimartingales," De Gruyter, Berlin, 1982.

Chapter 7

Laws of Large Numbers and Langevin Approximations for Stochastic Neural Field Equations

Martin G. Riedler · Evelyn Buckwar

Institute for Stochastics, Johannes Kepler University, Linz, Austria

ABSTRACT

In this study, we consider limit theorems for microscopic stochastic models of neural fields. We show that the Wilson–Cowan equation can be obtained as the limit in uniform convergence on compacts in probability for a sequence of microscopic models when the number of neuron populations distributed in space and the number of neurons per population tend to infinity. This result also allows to obtain limits for qualitatively different stochastic convergence concepts, e.g., convergence in the mean. Further, we present a central limit theorem for the martingale part of the microscopic models which, suitably re-scaled, converges to a centred Gaussian process with independent increments. These two results provide the basis for presenting the neural field Langevin equation, a stochastic differential equation taking values in a Hilbert space, which is the infinite-dimensional analogue of the chemical Langevin equation in the present setting. On a technical level, we apply recently developed law of large numbers and central limit theorems for piecewise deterministic processes taking values in Hilbert spaces to a master equation formulation of stochastic neuronal network models. These theorems are valid for processes taking values in Hilbert spaces, and by this are able to incorporate spatial structures of the underlying model.

KEYWORDS

Stochastic neural field equation Wilson–Cowan model Piecewise deterministic Markov process Stochastic processes in infinite dimensions Law of large numbers Martingale central limit theorem Chemical Langevin equation

1 INTRODUCTION

The present study is concerned with the derivation and justification of neural field equations from finite size stochastic particle models, i.e., stochastic models for the behaviour of individual neurons distributed in finitely many populations, in terms of mathematically precise probabilistic limit theorems. We illustrate this approach with the example of the Wilson–Cowan equation

$$\tau \dot{\nu}(t,x) = -\nu(t,x) + f\left(\int_D w(x,y)\nu(t,y)\,\mathrm{d}y + I(t,x)\right). \tag{1}$$

We focus on the following two aspects:

(A) Often one wants to study deterministic equations such as Eq. (1.1) in order to obtain results on the 'behaviour in the mean' of an intrinsically stochastic system. Thus, we first discuss limit theorems of the law of large numbers type for the limit of infinitely many particles. These theorems connect the trajectories of the stochastic particle models to the deterministic solution of mean field equations, and hence provide a justification studying Eq. (1.1) in order to infer on the behaviour of the stochastic system.

(B)Secondly, we aim to characterise the internal noise structure of the complex discrete stochastic models as in the limit of large numbers of neurons the noise is expected to be close to a simpler stochastic process. Ultimately, this yields a stochastic neural field model in terms of a stochastic evolution equation conceptually analogous to the Chemical Langevin Equation. The Chemical Langevin Equation is widely used in the study of chemical reactions networks for which the stochastic effects cannot be neglected but a numerical or analytical study of the exact discrete model is not possible due to its inherent complexity.

In this study, we understand as a microscopic model a description as a stochastic process, usually a Markov chain model, also called a master equation formulation (cf. [3, 5, 8, 9, 22] containing various master equation formulations of neural dynamics). In contrast, a macroscopic model is a deterministic evolution equation such as (1.1). Deterministic mean field equations have been used widely and for a long time to model and analyse large scale behaviour of the brain. In their original deterministic form, they are successfully used to model geometric visual hallucinations, orientation tuning in the visual cortex and wave propagation in cortical slices to mention only a few applications. We refer to [7] for a recent review and an extensive list of references. The derivation of these equations is based on a number of arguments from statistical physics and for a long time a justification from microscopic models has not been available. The interest in deriving mean field equations from stochastic microscopic model has been revived recently as it contains the possibility to derive deterministic 'corrections' to the mean field equations, also called second-order approximations. These corrections might account for the inherent stochasticity, and thus incorporate so called finite size effects. This has been achieved by either applying a path-integral approach to the master equation [8, 9] or by a van Kampen system-size expansion of the master equation [5]. In more detail, the author in the latter reference proposes a particular master equation for a finite number of neuron populations and derives the Wilson–Cowan equation as the first-order approximation to the mean via employing the van Kampen system size expansion and then taking the continuum limit for a continuum of populations. In keeping also the second-order terms, a 'stochastic' version of the mean field equation is also presented in the sense of coupling the first moment equation to an equation for the second moments.

However, the van Kampen system size expansion does not give a precise mathematical connection, as it neither quantifies the type of convergence (quality of the limit), states conditions when the convergence is valid nor does it allow to characterise the speed of convergence. Furthermore, particular care has to be taken in systems possessing multiple fixed points of the macroscopic equation, and we refer to [5] for a discussion of this aspect in the neural field setting. The limited applicability of the van Kampen system size expansion was already well known to Sect. 10 in van Kampen [33]. In parallel to the work of van Kampen, T. Kurtz derived precise limit theorems connecting sequences of continuous time Markov chains to solutions of systems of ordinary differential equations; see the seminal studies [19, 20] or the monograph [15]. Limit theorems of that type are usually called the fluid limit, thermodynamic limit, or hydrodynamic limit; for a review, see, e.g., [13].

As is thoroughly discussed in [5] establishing the connection between master equation models and mean field equations involves two limit procedures. First, a limit which takes the number of particles, in this case neurons per considered population, to infinity (thermodynamic limit), and a second which gives the mean field by taking the number of populations to infinity (continuum limit). In this 'double limit', the theorems by Kurtz describe the connection of taking the number of neurons per population to infinity yielding a system of ordinary differential equation, one for each population. Then the extension from finite to infinite dimensional state space is obtained by a continuum limit. This procedure corresponds to the approach in [5]. Thus, taking the double limit step by step raises the question what happens if we first take the spatial limit and then the fluid limit, thus reversing the order of the limit procedures, or in the case of taking the limits simultaneously. Recently, in an extension to the work of Kurtz, one of the present authors and co-authors established limit theorems that achieve this double limit [27], thus being able to connect directly finite population master equation formulations to spatio-temporal limit systems, e.g., partial differential equation or integro-differential equations such as the Wilson–Cowan equation (1.1). In a general framework, these limit theorems were derived for Piecewise Deterministic Markov Processes on Hilbert spaces, which in addition to the jump evolution also allow for a coupled deterministic continuous evolution. This generality was motivated by applications to neuron membrane models consisting of microscopic models of the ion channels coupled to a deterministic equation for the transmembrane potential. We find that this generality is also advantageous for the present situation of a pure jump model as it allows to include time-dependent inputs. In this study, we employ these theorems to achieve the aims (A) and (B) focussing on the example of the deterministic limit given by the Wilson–Cowan equation (1.1).

Finally, we state what this study does not contain, which in particular distinguishes the present study from [5, 8, 9] beyond mathematical technique. Presently, the aim is not to derive moment equations, i.e., a deterministic set of equations that approximate the moments of the Markovian particle model, but rather processes (deterministic or stochastic) to which a sequence of microscopic models converges under suitable conditions in a probabilistic way. This means that a microscopic model, which is close to the limit—presently corresponding to a large number of neurons in a large number of populations—can be assumed to be close to the limiting processes in structure and pathwise dynamics as indicated by the quality of the stochastic limit. Hence, the present work is conceptually—though neither in technique nor results—close to [30] wherein using a propagation to chaos approach in the vicinity of neural field equations the author also derives in a mathematically precise way a limiting process to finite particle models. However, it is an obvious consequence that the convergence of the models necessarily implies a close resemblance of their moment equations. This provides the connection to [5, 8, 9], which we briefly comment on in Appendix B.

As a guide, we close this introduction with an outline of the subsequent sections and some general remarks on the notation employed in this study. In Sects. 1.1 to 1.3, we first discuss the two types of mean field models in more detail, on the one hand, the Wilson–Cowan equation as the macroscopic limit and, on the other hand, a master equation formulation of a stochastic neural field. The main results of the paper are found in Sect. 2. There we set up the sequence of microscopic models and state conditions for convergence. Limit theorems of the law of large numbers type are presented in Theorem 2.1 and Theorem 2.2 in Sect. 2.1. The first is a classical weak law of large numbers providing uniform convergence on compacts in probability and the second convergence in the mean uniformly over the whole positive time axis. Next, a central limit theorem for the martingale part of the microscopic models is presented in Sect. 2.2 characterising the internal fluctuations of the model to be of a diffusive nature in the limit. This part of the study is concluded in Sect. 2.3 by presenting the Langevin approximations that arise as a result of the preceding limit theorems. The proofs of the theorems in Sect. 2 are deferred to Sect. 4. The study is concluded in Sect. 3 with a discussion of the implications of the presented results and an extension of these limit theorems to different master equation formulations or mean field equations.

Notations and Conventions Throughout the study, we denote by $L^p(D)$, $1 \leq p \leq \infty$, the Lebesgue spaces of real functions on a domain $D \subset \mathbb{R}^d$, $d \geq 1$. Physically reasonable choices are $d \in \{1, 2, 3\}$, however, for the mathematical theory presented the spatial dimension can be arbitrary. In the present study, spatial domains D are always bounded with a sufficiently smooth boundary, where the minimal assumption is a strong local Lipschitz condition; see [2]. For bounded domains D, this condition simply means that for every point on the boundary its neighbourhood on the boundary is the graph of a Lipschitz continuous function. Furthermore, for $\alpha \in \mathbb{N}$ we denote by $H^\alpha(D)$ the Sobolev spaces, i.e., subspaces of L 2 (D) L2(D)$L^2(D)$, with the corresponding Sobolev norm. For $\alpha \in \mathbb{R}_+ \setminus \mathbb{N}$ we denote by $H^\alpha(D)$ the interpolating Besov spaces. In this study, $H^{-\alpha}(D)$ is the dual space of $H^\alpha(D)$, which is in contrast to the widespread notation to denote by $H^{-\alpha}(D)$, α≥0 α≥0$\alpha \geq 0$, the dual space of $H_0^\alpha(D)$. As usual, we have $H^0(D) = L^2(D) = H^{-0}(D)$. We thus obtain a continuous scale of Hilbert spaces $H^\alpha(D)$, $\alpha \in \mathbb{R}$, which satisfy that $H^{\alpha_1}(D)$ is continuously embeddeda in $H^{\alpha_2}(D)$ for all α 1 <α 2 α1<α2$\alpha_1 < \alpha_2$. Next, a pairing $(\cdot, \cdot)_{H^\alpha}$ denotes the inner product of the Hilbert space $H^\alpha(D)$ and pairings in angle brackets $\langle \cdot, \cdot \rangle_{H^\alpha}$ denote the duality pairing for the Hilbert space $H^\alpha(D)$. That is, for $\psi \in H^\alpha(D)$ and $\phi \in H^{-\alpha}(D)$ the expression $\langle \phi, \psi \rangle_{H^\alpha}$ denotes the application of the real, linear functional ф to ψ. Furthermore, the spaces H α ($H^\alpha(D)$, $L^2(D)$ and $H^{-\alpha}(D)$ form an evolution triplet, i.e., the embeddings are dense and the application of linear functionals and the inner product in $L^2(D)$ satisfy the relation

$$\langle \phi, \psi \rangle_{H^\alpha} = (\phi, \psi)_{L^2} \quad \forall \phi \in L^2(D), \psi \in H^\alpha(D). \tag{1.2}$$

Norms in Hilbert spaces are denoted by $\|\cdot\|_{H^\alpha}$, $\|\cdot\|_0$ is used to denote the supremum norm of real functions, i.e., for $f : \mathbb{R} \to \mathbb{R}$ we have $\|f\|_0 = \sup_{z \in \mathbb{R}} |f(z)|$, and $|\cdot|$ denotes either the absolute value for scalars or the Lebesgue measure for measurable subsets of Euclidean space. Finally, we use $\mathbb{N}_0$ to denote the set of integers including zero.

1.1 The Macroscopic Limit

Neural field equations are usually classified into two types: rate-based and activity-based models. The prototype of the former is the Wilson–Cowan equation; see Eq. (1.1), which we also restate below, and the Amari equation, see Eq. (3.7) in Sect. 3, is the prototype of the latter. Besides being of a different structure, due to their derivation, the variable they describe has a completely different interpretation. In rate-based models, the variable describes the average rate of activity at a certain location and time, roughly corresponding to the fraction of active neurons at a certain infinitesimal area. In activity-based models, the macroscopic variable is an average electrical potential produced by neurons at a certain location. For a concise physical derivation that leads to these models, we refer to [5]. In the following, we consider rate-based equations, in particular, the classical Wilson–Cowan equation, to discuss the type of limit theorems we are able to obtain. We remark that the results are essentially analogous for activity based models.

Thus, the macroscopic model of interest is given by the equation

$$\tau \dot{v}(t,x) = -v(t,x) + f\left(\int_D w(x,y)v(t,y)\,\mathrm{d}y + I(t,x)\right), \tag{1.3}$$

where $\tau > 0$ is a decay time constant, $f : \mathbb{R} \to \mathbb{R}_+$ is a gain (or response) function that relates inputs that a neuron receives to activity. In (1.3), the value $f(z)$ can be interpreted as the fraction of neurons that receive at least threshold input. Furthermore, $w(x,y)$ is a weight function, which states the connectivity strength of a neuron located at y to a neuron located at x, and finally, $I(t,x)$ is an external input, which is received by a neuron at x at time t. For the weight function $w : D \times D \to \mathbb{R}$ and the external input I, we assume that $w \in L^2(D \times D)$ and $I \in C(\mathbb{R}_+, L^2(D))$. As for the gain function f, we assume in this study that f is non-negative, satisfies a global Lipschitz condition with constant $L > 0$, i.e.,

$$\left|f(a) - f(b)\right| \le L|a-b| \quad \forall a,b \in \mathbb{R}, \tag{1.4}$$

and it is bounded. From an interpretive point-of-view, it is reasonable and consistent to stipulate that f is bounded by one—being a fraction—as well as being monotone. The latter property corresponds to the fact that higher input results in higher activity. In specific models, f is often chosen to be a sigmoidal function, e.g., $f(z) = (1 + \mathrm{e}^{-(\beta_1 z + \beta_2)})^{-1}$ in [6] or $f(z) = (\tanh(\beta_1 z + \beta_2) + 1)/2$ in [3], which both satisfy $f \in [0,1]$. Moreover, the most common choices of f are even infinitely often differentiable with bounded derivatives, which already implies the Lipschitz condition (1.4).

The Wilson–Cowan equation (1.3) is well-posed in the strong sense as an integral equation in $L^2(D)$ under the above conditions. That is, Eq. (1.3) possesses a unique, continuously differentiable global solution v to every initial condition $v(0) = v_0 \in L^2(D)$, i.e., $v \in C^1([0,T], L^2(D))$ for all $T > 0$, which depends continuously on the initial condition. Furthermore, if the initial condition satisfies $v_0(x) \in [0, \|f\|_0]$ almost everywhere in D, then it holds for all $t > 0$ that $v(t,x) \in (0, \|f\|_0)$ for almost all $x \in D$. For a brief derivation of these results, we refer to Appendix A where we also state a result about higher spatial regularity of the solution: Let αEN αENα $\in \mathbb{N}$ be such that $\alpha > d/2$. If now $v_0 \in H^\alpha(D)$ and if f is at least α-times differentiable with bounded derivatives and the weights and the input function satisfy $w \in H^\alpha(D \times D)$ and $I \in C(\mathbb{R}_+, H^\alpha(D))$, then the equation is well-posed in $H^\alpha(D)$, i.e., for all $T > 0$ in $v \in C^1([0,T], H^\alpha(D))$. In particular, this implies that the solution v is jointly continuous on $\mathbb{R}_+ \times D$.

1.2 Master Equation Formulations of Neural Network Models

For the microscopic model, we concentrate on a variation of the model considered in [5, 6], which is already an improvement on a model introduced in [11]. We extend the model including variations among neuron populations and foremost time-dependent inputs. We chose this model over the master equation formulations in [8, 9] as it provides a more direct connection of the microscopic and macroscopic models; see also the discussion in Sect. 3. We describe the main ingredients of the model beginning with the simpler, time-independent model as prevalent in the literature. Subsequently, in Sect. 1.3 the final, time-dependent model is defined.

We denote by P the number of neuron populations in the model. Further, we assume that the k th neuron population consists of identical neurons which can either be in one of two possible states, active, i.e., emitting action potentials, and inactive, i.e., quiescent or not emitting action potentials. Transitions between states occur instantaneously and at random times. For all $k = 1, \ldots, P$, the random variables Θ_t^k denote the number of active neurons at time t. An integer $l(k)$ is used to characterise the population size. This number $l(k)$ can be interpreted as the number of neurons in the k th population, at least for sufficiently large values. However, this is not accurate in the literal sense as it is possible with positive probability for populations to contain more than $l(k)$ active neurons. Nevertheless, a posteriori the interpretation can be salvaged from the obtained limit theorems.b It is a corollary of these that the probability of more then $l(k)$ neurons being active for some time becomes arbitrarily small for large enough $l(k)$. Hence, for physiological reasonable neuron numbers the probability in these models of observing 'non-physiological' trajectories in the interpretation becomes ever smaller.

Proceeding with notation, $\Theta_t = (\Theta_t^1, \ldots, \Theta_t^P)$ is a (unbounded) piecewise constant stochastic process taking values in $\mathbb{N}_0^P$. The stochastic transitions from inactive to active states and vice versa for a neuron in population k are governed by a constant inactivation rate $\tau^{-1} > 0$ —uniformly for all populations—and inputs from other neurons depending on the current network state. This non-negative activation rate is given by $\tau^{-1} l(k) \overline{f}_k(\theta)$ for $\theta \in \mathbb{N}_0^P$. For the definition of $\overline{f}_k$, we consider weights $\overline{W}_{kj}$, $k, j = 1, \ldots, P$, which weigh the input one neuron in population k receives from a neuron in population j. Then the activation rate of a neuron in population k is proportional to

$$\overline{f}_k(\theta) = f\left(\sum_{j=1}^{P} \overline{W}_{kj}\theta^j\right) \tag{1.5}$$

for a non-negative function $f : \mathbb{R} \to \mathbb{R}$, which obviously corresponds to the gain function f in the Wilson–Cowan equation (1.3). We remark that here f is not the rate of activation of one neuron. In this model, the activation rate of a population is not proportional to the number of inactive neurons but it is proportional to $l(k)$ which stands for the total number of neurons in the population. In [5], this rate is thus interpreted as the rate with which a neuron becomes or remains active.

It follows that the process $(\Theta_t)_{t\geq 0}$ is a continuous-time Markov chain which is usually defined via the following master equation, where e k eke_k denotes the k th basis vector of $\mathbb{R}^P$,

$$\begin{aligned}\frac{\mathrm{d}\mathbb{P}[\theta,t]}{\mathrm{d}t} = \frac{1}{\tau}\sum_{k=1}^{P}\Big(&l(k)\overline{f}_k(\theta - e_k)\mathbb{P}[\theta - e_k, t] \\ &- \left(\theta^k + l(k)\overline{f}_k(\theta)\right)\mathbb{P}[\theta, t] + \left(\theta^k + 1\right)\mathbb{P}[\theta + e_k, t]\Big)\end{aligned} \tag{1.6}$$

which is endowed with the boundary conditions $\mathbb{P}[\theta, t] = 0$ if $\theta \notin \mathbb{N}_0^P$. In (1.6), the variable P[θ,t] P[θ,t]$\mathbb{P}[\theta, t]$ denotes the probability that the process Θ_t is in state θ at time t. Finally, the definition is completed with stating an initial law $\mathcal{L}$, the distribution of Θ_0, i.e., providing an initial value for the ODE system (1.6).

Another definition of a continuous-time Markov chain is via its generator; see, e.g., [15]. Although the master equation is widely used in the physics and chemical reactions literature the mathematically more appropriate object for the study of a Markov process is its generator and the master equation is an object derived from the generator, see Sect. V in [33]. The generator of a Markov process is an operator defined on the space of real functions over the state space of the process. For the above model defined by the master equation (1.6), the generator is given by

$$\mathcal{A}g(\theta) = \lambda(\theta) \int_{\mathbb{N}_0^P} \big(g(\xi) - g(\theta)\big)\mu(\theta, \mathrm{d}\xi) \tag{1.7}$$

for all suitable $g : \mathbb{N}_0^P \to \mathbb{R}$. For details, we refer to [15]. Here, λ is the total instantaneous jump rate, given by

$$\lambda(\theta) := \frac{1}{\tau} \sum_{k=1}^{P} \big(\theta^k + I(k)\overline{f}_k(\theta)\big). \tag{1.8}$$

and defines the distribution of the waiting time until the next jump, i.e.,

$$\mathbb{P}\big[\Theta_{t+s} = \Theta_t \ \forall s \in [0, \Delta t] | \Theta_t = \theta\big] = \mathrm{e}^{-\lambda(\theta)\Delta t}.$$

Further, the measure μ in (1.7) is a Markov kernel on the state space of the process defining the conditional distribution of the post-jump value, i.e.,

$$\mathbb{P}[\Theta_t \in A | \Theta_t \neq \Theta_{t-}] = \mu(\Theta_{t-}, A) \tag{1.9}$$

for all sets $A \subseteq \mathbb{N}_0^P$. In the present case for each θ, the measure μ is given by the discrete distribution

$$\mu\big(\theta, \{\theta - e_k\}\big) = \frac{1}{\tau} \frac{\theta^k}{\lambda(\theta)},$$

$$\mu\big(\theta, \{\theta + e_k\}\big) = \frac{1}{\tau} \frac{I(k)\overline{f}_k(\theta)}{\lambda(\theta)} \quad \forall k = 1, \ldots, P. \tag{1.10}$$

The importance of the generator lies in the fact that it fully characterises a Markov process and that convergence of Markov processes is strongly connected to the convergence of their generators; see [15].

1.3 Including External Time-Dependent Input

Until now, the microscopic model does not incorporate any time-dependent input into the system. In analogy to the macroscopic equation (1.3), this input enters into the model inside the active rate function $\overline{f}_k$. Thus, let $\overline{I}_k(t)$ denote the external input into a neuron in population k at time t, then the time-dependent activation rate is given by

$$\overline{f}_k(\theta, t) = f\left(\sum_{j=1}^{P} \overline{W}_{kj}\theta^j + \overline{I}_k(t)\right). \tag{1.11}$$

The most important qualitative difference when substituting (1.5) by (1.11) is that the corresponding Markov process is no longer homogeneous. In particular, the waiting time distributions in between jumps are no longer exponential, but satisfy

$$\mathbb{P}\left[\Theta_{t+s} = \Theta_t \ \forall s \in [0, \Delta t] | \Theta_t = \theta\right] = \mathrm{e}^{-\int_0^{\Delta t} \lambda(\theta, s)\,\mathrm{d}s}.$$

Hence, the resulting process is an inhomogeneous continuous-time Markov chain; see, e.g., Sect. 2 in [36]. It is straightforward to write down the corresponding master equation analogously to (1.6) yielding a system of non-autonomous ordinary differential equations, cf. the master equation formulation in [8]. Similarly, there exists the notion of a time-dependent generator for inhomogeneous Markov processes, cf. Sect. 4.7 in [15]. Employing a standard trick, that is, suitably extending the state space of the process, we can transform a inhomogeneous to a homogeneous Markov process [15, 28]. That is, the space-time process $Y_t := (\Theta_t, t)$ is again a homogeneous Markov process. The initial law of the associated space-time process is $\mathcal{L} \times \delta_0$ on $\mathbb{N}^P \times \mathbb{R}_+$. We emphasise that definitions of the space-time process and its initial law imply that the time-component starts at 0 a.s. and, moreover, moves continuously and deterministically. That is, the trajectories satisfy in between jumps the differential equation

$$\begin{pmatrix} \dot{\theta} \\ \dot{t} \end{pmatrix} = \begin{pmatrix} 0 \\ 1 \end{pmatrix},$$

where the jump intensity λ is given by the sum of all individual time-dependent rates analogously to (1.8). Finally, the post jump value is given by a Markov kernel $\mu((\theta, t), \cdot) \times \delta_t$ as there clearly do not occur jumps in the progression of time and μ is the obvious time-dependent modification of (1.10).

It thus follows, that the space-time process $(\Theta_t, t)_{t\geq 0}$ is a homogeneous Piecewise Deterministic Markov Process (PDMP); see, e.g., [14, 16, 26]. This connection is particularly important as we apply in the course of the present study limit theorems developed for this type of processes; see [27]. Finally, for the space-time process $(\Theta_t, t)_{t\geq 0}$, we obtain for suitable functions $g : \mathbb{N}_0^P \times \mathbb{R}_+ \to \mathbb{R}$ the generator

$$\mathcal{A}g(\theta,t) = \partial_t g(\theta,t) + \lambda(\theta,t)\int_{\mathbb{N}_0^P}\big(g(\xi,t) - g(\theta,t)\big)\mu\big((\theta,t),\mathrm{d}\xi\big). \tag{1.12}$$

2 A PRECISE FORMULATION OF THE LIMIT THEOREMS

In this section, we present the precise formulations of the limit theorems. To this end, we first define a suitable sequence of microscopic models, which gives the connection between the defining objects of the Wilson–Cowan equation (1.3) and the microscopic models discussed in Sect. 1.2. Thus, $(Y_t^n)_{t\geq 0} = (\Theta_t^n, t)_{t\geq 0}$, $n \in \mathbb{N}$, denotes a sequence of microscopic PDMP neural field models of the type as defined in Sect. 1.3. Each process $(Y_t^n)_{t\geq 0}$ is defined on a filtered probability space $(\Omega^n, \mathcal{F}^n, (\mathcal{F}_t^n)_{t\geq 0}, \mathrm{P}^n)$, which satisfies the usual conditions. Hence, the defining objects for the jump models are now dependent on an additional index n. That is $P(n)$ denotes the number of neuron populations in the n th model, $l(k,n)$ is the number of neurons in the k th population of the n th model and analogously we use the notations $\overline{W}_{kj}^n$ and $\overline{I}_{k,n}$ and $\overline{f}_{k,n}$. However, we note from the beginning that the decay rate τ^{-1} is independent of n and τ is the time constant in the Wilson–Cowan equation (1.3). In the following paragraphs, we discuss the connection of the defining components of this sequence of microscopic models to the components of the macroscopic limit.

Connection to the Spatial Domain D A key step of connecting the microscopic models to the solution of Eq. (1.3) is that we need to put the individual neuron populations into relation to the spatial domain D the solution of (1.3) lives on. To this end, we assume that each population is located within a sub-domain of D and that the sub-domains of the individual populations are non-overlapping. Hence, for each $n \in \mathbb{N}$, we obtain a collection $\mathcal{D}_n$ of $P(n)$ non-overlapping sub-sets of D denoted by $D_{1,n}, \ldots, D_{P(n),n}$. We assume that each subdomain is measurable and convex. The convexity of the sub-domains is a technical condition that allows us to apply Poincaré's inequality, cf. (4.1). We do not think that this condition is too restrictive as most reasonable partition domains, e.g., cubes, triangles, are convex. Furthermore, for all reasonable domains D, e.g., all Jordan measurable domains, a sequence of convex partitions can be found such that additionally the conditions imposed in the limit theorems below are also satisfied. One may think of obtaining the collection $\mathcal{D}_n$ by partitioning the domain into $P(n)$ convex sub-domains $D_{1,n}, \ldots, D_{P(n),n}$ and confining each neuron population to one sub-domain. However, it is not required that the union of the sets in $\mathcal{D}_n$ amounts to the full domain D nor that the partitions consists of refinements. Necessary conditions on the limiting behaviour of the sub-domains are very strongly connected to the convergence of initial conditions of the models, which is a condition in the limit theorems; see below. For the sake of terminological simplicity, we refer to $\mathcal{D}_n$ simply as the partitions.

We now define some notation for parameters characterising the partitions D n Dn$\mathcal{D}_n$: the minimum and maximum Lebesgue measure, i.e., length, area, or volume depending on the spatial dimension, is denoted by

$$v_-(n) := \min_{k=1,\ldots,P(n)} |D_{k,n}|, \qquad v_+(n) := \max_{k=1,\ldots,P(n)} |D_{k,n}|. \tag{2.1}$$

$$\delta_+(n) := \max_{1,\ldots,P(n)} \operatorname{diam}(D_{k,n}), \tag{2.2}$$

where the diameter of a set $D_{k,n}$ is defined as diam$\operatorname{diam}(D_{k,n}) := \sup_{x,y\in D_{k,n}} |x-y|$. In the special case of domains obtained by unions of cubes with edge length n^{-1}, it obviously holds that $v_\pm(n) = n^{-d}$ and $\delta_+(n) = \sqrt{d}n^{-1}$. It is a necessary condition in all the subsequent limit theorems that lim $\lim_{n\to\infty} \delta_+(n) = 0$. This condition implies on the one hand that lim $\lim_{n\to\infty} v_+(n) = 0$ as the Lebesgue measure of a set is bounded in terms of its diameter, and on the other hand—at least in all but degenerate cases due to the necessary convergence of initial conditions that $\lim_{n\to\infty} P(n) = \infty$. That is, in order to obtain a limit the sequence of partitions usually consists of ever finer sets and the number of populations diverges. Finally, each domain $D_{k,n}$ of the partition $\mathcal{D}_n$ contains one neuron population 'consisting' of $l(k,n) \in \mathbb{N}$ neurons. Then we denote by $\ell_\pm(n)$ the maximum and minimum number of neurons in populations corresponding to the n th model, i.e.,

$$\ell_-(n) := \min_{k=1,\ldots,P(n)} l(k,n), \qquad \ell_+(n) := \max_{k=1,\ldots,P(n)} l(k,n). \tag{2.3}$$

Connection to the Weight Function w We assume that there exists a function $w : D \times D \to \mathbb{R}$ such that the connection to the discrete weights is given by

$$\overline{W}^n_{kj} := \frac{1}{|D_{k,n}|} \int_{D_{k,n}} \left(\int_{D_{j,n}} w(x,y)\,\mathrm{d}y \right) \mathrm{d}x, \tag{2.4}$$

where w is the same function as in the Wilson–Cowan equation (1.3). For the definition of activation rate at time t, we thus obtain

$$\overline{f}_{k,n}(\theta^n, t) := f\left(\sum_{j=1}^{P} \overline{W}^n_{kj} \frac{\theta^{j,n}}{l(j,n)} + \overline{I}_{k,n}(t) \right). \tag{2.5}$$

As already highlighted by Bressloff [5], the transition rates are not uniquely defined by the requirement that a possible limit to the microscopic models is given by the Wilson–Cowan equation (1.3). If in (2.5), the definition of the transition rates is changed to

$$\overline{f}_{k,n}(\theta^n, t) := f^n\left(\sum_{j=1}^{P} \overline{W}^n_{kj} \frac{\theta^{j,n}}{l(j,n)} + \overline{I}_{k,n}(t)\right).$$

where $f^n, n \in \mathbb{N}$, is a sequence of functions converging uniformly to f, then all limit theorems remain valid. The proof can be carried out as presented adding and subtracting the appropriate term where the additional difference term vanishes due to $\sup_{x\in\mathbb{R}} |f^n(x) - f(x)| \to 0$ for $n \to \infty$. Hence, any microscopic model with gain rates $\overline{f}_{k,n}$ of such a form reduces to the same Wilson–Cowan equation in the limit. Clearly, the same applies analogously to the decay rate τ, the weights w, and the input I.

Connection to the Input Current I

The external input which is applied to neurons in a certain population is obtained by spatially averaging a space-time input over the sub-domain that population is located in, i.e.,

$$\overline{I}_{k,n}(t) := \frac{1}{|D_{k,n}|} \int_{D_{k,n}} I(t, x)\, dx. \tag{2.6}$$

This completes the definition of the Markov jump processes $(\Theta^n_t, t)_{t\geq 0}$. For the

$$\lambda^n(\theta^n, t) := \frac{1}{\tau} \sum_{k=1}^{P} \left(\theta^{k,n} + l(k,n)\overline{f}_{k,n}(\theta^n, t)\right)$$

and the transition measure μ n μnμ^n is defined by

$$\mu^n\left((\theta^n, t), \{\theta^n - e_k\}\right) := \frac{1}{\tau} \frac{\theta^{k,n}}{\lambda^n(\theta^n, t)},$$

$$\mu^n\left((\theta^n, t), \{\theta^n + e_k\}\right) := \frac{1}{\tau} \frac{l(k,n)\overline{f}_{k,n}(\theta^n, t)}{\lambda^n(\theta^n, t)}$$

for all $k = 1, \ldots, P(n)$.

Connection to the Solution v As functions of time, the paths of the PDMP $(\Theta^n_t, t)_{t\geq 0}$ and the solution v live on different state spaces. The former takes values in $\mathbb{N}_0^P \times \mathbb{R}_+$ and the latter in $L^2(D)$. Thus, in order to compare these two, we have to introduce a mapping that maps the stochastic process onto $L^2(D)$. In [27], the authors called such a mapping a coordinate function, which is also the terminology used in [13]. In fact, the limit theorems we subsequently present actually are for the processes we obtain from the composition of the coordinate functions with the PDMPs. Here, it is important to note that for each $n \in \mathbb{N}$ the coordinate functions may—and usually do—differ, however, they project the process into the common space $L^2(D)$. For the mean field models, we define the coordinate functions for all $n \in \mathbb{N}$ by

$$\nu^n : \mathbb{N}_0^P \to L^2(D) : \theta^n \mapsto \sum_{k=1}^{P} \frac{\theta^{k,n}}{l(k,n)} \mathbb{I}_{D_{k,n}}. \tag{2.7}$$

Clearly, each ν^n is a measurable map into $L^2(D)$. For the composition of ν^n with the stochastic process $(\Theta_t^n, t)_{t \geq 0}$, we also use the abbreviation $\nu_t^n := \nu^n(\Theta_t^n)$, and hence the resulting stochastic process $(\nu_t^n)_{t\geq 0}$ is an adapted càdlàg process taking values in $L^2(D)$. This process thus states the activity at a location $x \in D$ as the fraction of active neurons in the population, which is located around this location.

Connection of the Initial Conditions

One condition in the subsequent limit theorems is the convergence of initial conditions in probability, i.e., the assumption that

$$\lim_{n\to\infty} \mathbb{P}^n\left[\left\| \nu^n\left(\Theta_0^n\right) - \nu_0 \right\|_{L^2(D)} > \epsilon\right] = 0 \quad \forall \epsilon > 0. \tag{2.8}$$

It is easy to see that such a sequence of initial conditions Θ n 0 Θ0nΘ_0^n, nEN nEN$n \in \mathbb{N}$, can be found for any deterministic initial condition v 0 v0ν_0 under some reasonable conditions on the domain D and the sequence of partitions D n Dn$\mathcal{D}_n$. Hence, the assumption (2.8) can always be satisfied. For example, we may define such a sequence of initial conditions by

$$\Theta_0^{k,n} = \operatorname*{argmin}_{i=1,\ldots,l(k,n)} \left| \frac{i}{l(k,n)} - \frac{1}{|D_{k,n}|} \int_{D_{k,n}} \nu(0,x)\,\mathrm{d}x \right|.$$

Next, assuming that partitions fill the whole domain D for $n \to \infty$, i.e., $\lim_{n\to\infty} |D \setminus \bigcup_{k=1}^{P(n)} D_{k,n}| = 0$, and that the maximal diameter of the sets decreases to zero, i.e.,$\lim_{n\to\infty} \delta_+(n) = 0$, it is easy to see using the Poincaré inequality (4.1) that the above definition of the initial condition implies that $\|\nu_0^n - \nu(0)\|_{L^2(D)} \to 0$ and $\sup_{n\in\mathbb{N}} \|\nu_0^n\|_{L^2(D)}^{2r} < \infty$ for all $r \geq 1$. Then (2.8) holds trivially as the initial condition is deterministic and converges. A simple non-degenerate sequence of initial conditions is obtained by choosing random initial conditions with the above value as their mean and sufficiently fast decreasing fluctuations. Furthermore, a sequence of partitions, which satisfy the above conditions also exists for a large class of reasonable domains D. Assume that D is Jordan measurable, i.e., a bounded domain such that the boundary is a Lebesgue null set, and let C_n be the smallest grid of cubes with edge length 1/n 1/n$^{1/n}$ covering D. We define D_n to be the set of all cubes, which are fully in D. As D is Jordan measurable, these partitions fill up D from inside and $\delta_+(n) \to 0$. For a more detailed discussion of these aspects, we refer to [26].

In the remainder of this section, we now collect the main results of this article. We start with the law of large numbers, which establishes the connection to the deterministic mean field equation, and then proceed to central limit theorems which provide the basis for a Langevin approximation. The proofs of the results are deferred to Sect. 4.

2.1 A Law of Large Numbers

The first law of large numbers takes the following form. Note that the assumptions imply that the number of neuron populations diverges.

Theorem 2.1 (Law of large numbers)

Let $w \in L^2(D) \times L^2(D)$ and $I \in L^2_{\rm loc}(\mathbb{R}_+, H^1(D))$. Assume that the sequence of initial conditions converges to $\nu(0)$ in probability in the space $L^2(D)$, i.e., (2.8) holds, that $\mathbb{E}^n \Theta_0^{k,n} \le l(k,n)$, and that

$$\lim_{n\to\infty} \delta_+(n) = 0, \qquad \lim_{n\to\infty} \ell_-(n) = \infty \tag{2.9}$$

holds. Then it follows that the sequence of $L^2(D)$ -valued jump-process $(\nu_t^n)_{t\ge 0}$ converges uniformly on compact time intervals in probability to the solution v of the Wilson–Cowan equation (1.3), i.e., for all $T, \epsilon > 0$ it holds that

$$\lim_{n\to\infty} \mathbb{P}^n\Big[\sup_{t\in[0,T]} \big\|\nu_t^n - \nu(t)\big\|_{L^2(D)} > \epsilon\Big] = 0. \tag{2.10}$$

Moreover, if for $r \ge 1$ the initial conditions satisfy in addition $\sup_{n\in\mathbb{N}} \mathbb{E}^n \|\nu_0^n\|^{2r}_{L^2(D)} < \infty$, then convergence in the rth mean holds, i.e., for all $T > 0$

$$\lim_{n\to\infty} \mathbb{E}^n \sup_{t\in[0,T]} \big\|\nu_t^n - \nu(t)\big\|^r_{L^2(D)} = 0. \tag{2.11}$$

Remark 2.1 The norm of the uniform convergence $\sup_{t\in[0,T]}\|\cdot\|_{L^2(D)}$, which we used in Theorem 2.1 is a very strong norm on the space of $L^2(D)$ -valued càdlàg functions on $[0,T]$. Hence, due to continuous embeddings, the result immediately extends to weaker norms, e.g., the norms $L^p((0,T),L^2(D))$ for all $1\le p\le\infty$. Also, for the state space, weaker spatial norms can be chosen, e.g., $L^p(D)$ with $1\le p\le 2$ or any norm on the duals $H^{-\alpha}(D)$ of Sobolev spaces with $\alpha>0$. If weaker norms for the state space are considered, it is possible to relax the conditions of Theorem 2.1 by sharpening some estimates in the proof of the theorem. The results in the following corollary cover the whole range of α≥0 α≥0$\alpha\ge 0$ and splits it into sections with weakening conditions. In particular note that after passing to weaker norms, the convergence does not necessitate that the neuron numbers per population diverge. However, regarding the divergence of the neuron populations, this condition $\delta_+(n)\to 0$) cannot be relaxed.

Corollary 2.1 Let $\alpha\ge 0$ and set

$$q:=\begin{cases}\frac{2d}{d+2\alpha} & if\, 0\le\alpha<d/2,\\ 1- & if\,\alpha=d/2,\\ 1 & if\, d/2<\alpha<\infty.\end{cases} \tag{2.12}$$

Further, assume that $w\in L^q(D)\times L^2(D)$ and $I\in L^2_{\text{loc}}(\mathbb{R}_+,H^1(D))$ and that the sequence of initial conditions converges to $\nu(0)$ in probability in the space $H^{-\alpha}(D)$, that $\lim_{n\to\infty}\delta_+(n)=0$ and

$$\left.\begin{array}{lll}\lim_{n\to\infty}\frac{v_+(n)^{2\alpha/d}}{\ell_-(n)}=0 & if\, 0\le\alpha<d/2,\\ \lim_{n\to\infty}\frac{v_+(n)^{1-}}{\ell_-(n)}=0 & if\,\alpha=d/2,\\ \lim_{n\to\infty}\frac{v_+(n)}{\ell_-(n)}=0 & if\, d/2<\alpha<\infty.\end{array}\right\} \tag{2.13}$$

where 1− denotes an arbitrary positive number strictly smaller than 1. Then it holds for all $T,\epsilon>0$ that

$$\lim_{n\to\infty}\mathbb{P}^n\Big[\sup_{t\in[0,T]}\|\nu_t^n-\nu(t)\|_{H^{-\alpha}(D)}>\epsilon\Big]=0$$

and for $r\ge 1$, if the additional boundedness assumptions of Theorem 2.1 are satisfied, that for all $T>0$

$$\lim_{n\to\infty}\mathbb{E}^n\sup_{t\in[0,T]}\|\nu_t^n-\nu(t)\|^r_{H^{-\alpha}(D)}=0.$$

Remark 2.2 We believe that fruitful and illustrative comparisons of these convergence results and their conditions to the results in Kotelenez [17, 18], and particularly, Blount [4] can be made. Here, we just mention that the latter author conjectured the conditions (2.13) to be optimal for the convergence, but was not able to prove this result in his model of chemical reactions with diffusions for the region α∈(0,d/2] α∈(0,d/2]$\alpha \in (0, d/2]$. For our model, we could achieve these rates.

2.1.1 Infinite-Time Convergence

In the law of large numbers, Theorem 2.1, and its Corollary 2.1 we have presented results of convergence over finite time intervals. Employing a different technique, we are also able to derive a convergence result over the whole positive time axis motivated by a similar result in [32]. The proof of the following theorem is deferred to Sect. 4.3. Restricted to finite time intervals, the subsequent result is strictly weaker than Theorem 2.1. However, the result is important when one wants to analyse the mean long time behaviour of the stochastic model via a bifurcation analysis of the deterministic limit as (2.14) suggests that $\mathbb{E}^n \nu_t^n$ is close to $\nu(t)$ for all times $t \geq 0$ for sufficiently large n.

Theorem 2.2 Let α≥0 α≥0$\alpha \geq 0$ and assume that the conditions of Corollary 2.1 are satisfied. We further assume that the current input function $I \in L^2_{\mathrm{loc}}(\mathbb{R}_+, H^1(D))$ satisfies $\|\nabla_x I\|_{L^\infty(\mathbb{R}_+, L^2(D))} < \infty$, i.e., it is square integrable in $H^1(D)$ over bounded intervals, and possesses first spatial derivatives bounded for almost all $t \geq 0$ in $L^2(D)$. Then it holds that

$$\lim_{n\to\infty} \sup_{t\geq 0} \mathbb{E}^n \left\| \nu_t^n - \nu(t) \right\|_{H^{-\alpha}(D)} = 0. \tag{2.14}$$

2.2 A Martingale Central Limit Theorem

In this section, we present a central limit theorem for a sequence of martingales associated with the jump processes v n vnν^n . A brief, heuristic discussion of the method of proof for the law of large numbers explains the importance of these martingales and motivates their study. In the proof of the law of large numbers, the central argument relies on the fact that the process $(\nu_t^n)_{t\geq 0}$ satisfies the decomposition

$$\nu_t^n = \nu_0^n + \int_0^t \lambda(\Theta_s^n, s) \int_{\mathbb{N}_0^P} (\nu^n(\xi) - \nu^n(\Theta_s^n))\mu^n((\Theta_s^n, s), \mathrm{d}\xi)\,\mathrm{d}s + M_t^n. \tag{2.15}$$

Here, the process $(M_t^n)_{t\geq 0}$ is a Hilbert space-valued, square-integrable, càdlàg martingale using (2.15) as its definition. We have used this representation of the process ν^n in the proof of Theorem 2.2; see Sect. 4.3. We note that the Bochner integral in (2.15) is a.s. well defined due to bounded second moments of the integrand; see (4.7) in the proof of Theorem 2.1. Now an heuristic argument to obtain the convergence to the solution of the Wilson–Cowan equation is the following: The initial conditions converge, the martingale term M^n converges to zero and the integral term in the right-hand side of (2.15) converges to the right-hand side in the Wilson–Cowan equation (1.3). Hence, the 'solution' ν^n of (2.15) converges to the solution v of the Wilson–Cowan equation (1.3). Now interpreting Eq. (2.15) as a stochastic evolution equation, which is driven by the martingale $(M_t^n)_{t\geq 0}$ sheds light on the importance of the study of this term. Because, from this point of view, the martingale part in the decomposition (2.15) contains all the stochasticity inherent in the system. Then the idea for deriving a Langevin or linear noise approximation is to find a stochastic non-trivial limit (in distribution) for the sequence of martingales and substituting heuristically this limiting martingale into the stochastic evolution equation. Then it is expected that this new and much less complex process behaves similarly to the process $(\nu_t^n)_{t\geq 0}$ for sufficiently large n. Deriving a suitable limit for $(M_t^n)_{t\geq 0}$ is what we set to do next. The result can be found in Theorem 2.3 below and takes the form of a central limit theorem.

First of all, what has been said so far implies the necessity of re-scaling the martingale with a diverging sequence in order to obtain a non-trivial limit. The conditions in the law of large numbers imply in particular that the martingale converges uniformly in the mean square to zero, i.e.,

$$\lim_{n\to\infty} \mathbb{E}^n \sup_{t\in[0,T]} \left\| M_t^n \right\|_{L^2(D)} = 0,$$

which in turn implies convergence in probability and convergence in distribution to the zero limit.

Furthermore, in contrast to Euclidean spaces norms on infinite-dimensional spaces are usually not equivalent. In Corollary 2.1, we exploited this fact as it allowed us to obtain convergence results under less restrictive conditions by changing to strictly weaker norms. In the formulation and proof of central limit theorems, the change to weaker norms even becomes an essential ingredient. It is often observed in the literature, see, e.g., [4, 17, 18] that central limit theorems cannot be proven in the strongest norm for which the law of large numbers holds, e.g., L 2 (D) L2(D)$L^2(D)$ in the present setting, but only in a strictly weaker norm. Here, this norm is the norm in the dual of an appropriate Sobolev space. Hence, from now on, we consider for all $n \in \mathbb{N}$ the processes $(\nu_t^n)_{t\geq 0}$ and the martingales $(M_t^n)_{t\geq 0}$ as taking values in the space $H^{-\alpha}(D)$ for an $\alpha > d$, where d is the dimension of the spatial domain D, using the embedding of $L^2(D)$ into $H^{-\alpha}(D)$. The technical significance of the restriction α>d α>d$\alpha > d$ is that these are the indices such that there exists an embedding $H^{\alpha}(D)$ into a $H^{\alpha_1}(D)$ with $d/2 < \alpha_1 < \alpha$, which is of Hilbert–Schmidt typec due to Maurin's theorem and $H^{\alpha_1}(D)$ is embedded into $C(\overline{D})$ due to the Sobolev embedding theorem. These two properties are essential for the proof of the central limit theorem and their occurrence will be made clear subsequently.

The limit we propose for the re-scaled martingale sequence is a centred diffusion process in $H^{-\alpha}(D)$, that is, a centred continuous Gaussian stochastic process $(X_t)_{t\geq 0}$ taking values in $H^{-\alpha}(D)$ with independent increments and given covariance $C(t)$, $t \geq 0$; see, e.g., [12, 25] for a discussion of Gaussian processes in Hilbert spaces. Such a process is uniquely defined by its covariance operator and conversely, each family of linear, bounded operators $C(t) : H^{\alpha}(D) \to H^{-\alpha}(D)$, t≥0 t≥0$t \geq 0$, uniquely defines a diffusion processd if

(i) each $C(t)$ is symmetric and positive, i.e.,

$$\langle C(t)\phi, \psi\rangle_{H^{\alpha}(D)} = \langle C(t)\psi, \phi\rangle_{H^{\alpha}(D)} \quad \text{and} \quad \langle C(t)\phi, \phi\rangle_{H^{\alpha}(D)} \geq 0.$$

(ii)each $C(t)$ is of trace class, i.e., for one (and thus every) orthonormal basis ϕφ_j, $j \in \mathbb{N}$, in $H^{\alpha}(D)$ it holds that

$$\sum_{j=1}^{\infty} \langle C(t)\varphi_j, \varphi_j\rangle_{H^{\alpha}(D)} < \infty. \tag{2.16}$$

(iii) and the family $C(t)$, t≥0 t≥0$t \geq 0$, is continuously increasing in t in the sense that the map $t \mapsto \langle C(t)\phi, \psi\rangle_{H^{\alpha}(D)}$ is continuous and increasing for all $\phi, \psi \in H^{\alpha}(D)$.

We next define the process, which will be the limit identified in the martingale central limit theorem via its covariance. In order to define the operator C, we first define a family of linear operators $G(v(t), t)$ mapping from $H^{\alpha}(D)$ into the dual space $H^{-\alpha}(D)$ via the bilinear form

$$\begin{aligned} &\langle G(v(t), t)\phi, \psi\rangle_{H^{\alpha}(D)} \\ &= \int_D \phi(x)\Big(\frac{1}{\tau} v(t, x) \\ &\quad + \frac{1}{\tau} f\Big(\int_D w(x, y) v(t, y)\,dy + I(t, x)\Big)\Big)\psi(x)\,dx. \end{aligned} \tag{2.17}$$

It is obvious that this bilinear form is symmetric and positive and, as v(t) v(t)$\nu(t)$ is continuous in t, it holds that the map $t \mapsto \langle G(\nu(t), t)\phi, \psi\rangle_{H^\alpha(D)}$ is continuous for all $\phi, \psi \in H^\alpha(D)$. Furthermore, it is easy to see that the operator is bounded, i.e.,

$$\left\| G(\nu(t), t) \right\|_{L(H^\alpha(D), H^{-\alpha}(D))} = \sup_{\|\phi\|_{H^\alpha(D)}=1} \sup_{\|\psi\|_{H^\alpha(D)}=1} \left| \left\langle G(\nu(t), t)\phi, \psi \right\rangle_{H^\alpha(D)} \right| < \infty,$$

as the solution of the Wilson–Cowan equation v and the gain function f are pointwise bounded. Hence, due to the Cauchy–Schwarz inequality, the norm $|\langle G(\nu(t), t)\phi, \psi\rangle_{H^\alpha(D)}|$ is proportional to the product $\|\phi\|_{L^2(D)} \|\psi\|_{L^2(D)}$ and for any $\alpha \geq 0$ the Sobolev embedding theorem gives now a uniform bound in terms of the norm of ϕ, ψ in $H^\alpha(D)$. As a final property, we show that these operators are of trace-class if $\alpha > d/2$. Thus, let $(\varphi_j)_{j\in\mathbb{N}}$ be an orthonormal basis in $H^\alpha(D)$, then the Cauchy–Schwarz inequality yields

$$\left| \left\langle G(\nu(t), t)\varphi_j, \varphi_j \right\rangle_{H^\alpha(D)} \right| \leq \frac{1}{\tau}(1 + \|f\|_0)|D| \|\varphi_j\|^2_{L^2(D)}.$$

Summing these inequalities for all $j \in \mathbb{N}$, we find that the resulting right-hand side is finite as due to Maurin's theorem the embedding of $H^\alpha(D)$ into $L^2(D)$ is of Hilbert–Schmidt type. Moreover, their trace is even bounded independently of t.

Now, it holds that the map $t \mapsto G(\nu(t), t)$ is continuous taking values in the Banach space of trace class operators, hence we define trace class operators $C(t)$ from $H^\alpha(D)$ into $H^{-\alpha}(D)$ via the Bochner integral for all t≥0 t≥0$t \geq 0$

$$C(t) := \int_0^t G(\nu(s), s)\, \mathrm{d}s. \tag{2.18}$$

Clearly, the resulting bilinear form $\langle C(t)\cdot, \cdot\rangle_{H^\alpha(D)}$ inherits the properties of the bilinear form (2.17). Moreover, due to the positivity of the integrands, it follows that $\langle C(t)\phi, \phi\rangle_{H^\alpha(D)}$ is increasing in t for all $\phi \in H^\alpha(D)$. Hence, the family of operators $C(t)$, $t \geq 0$, satisfies the above conditions (i)–(iii), and thus uniquely defines an $H^{-\alpha}(D)$ -valued diffusion process.

We are now able to state the martingale central limit theorem. The proof of the theorem is deferred to Sect. 4.4.

Theorem 2.3 (Martingale central limit theorem)

Let $\alpha > d$ and assume that the conditions of Theorem 2.1 are satisfied. In particular, convergence in the mean holds, i.e., (2.11) holds for r=1 r=1$r = 1$. Additionally, we assume it holds that

$$\lim_{n\to\infty} \frac{v_-(n)}{v_+(n)} \frac{\ell_-(n)}{\ell_+(n)} = 1.$$

(2.19)

Then it follows that the sequence of re-scaled $H^{-\alpha}(D)$ -valued martingales

$$\left(\sqrt{\frac{\ell_-(n)}{v_+(n)}} M_t^n\right)_{t\geq 0}$$

converges weakly on the space of $H^{-\alpha}(D)$ -valued càdlàg function to the $H^{-\alpha}(D)$ -valued diffusion process defined by the covariance operator $C(t)$ given by (2.18).

Remark 2.3 In connection with the results of Theorem 2.3, two questions may arise. First, in what sense is there uniqueness of the re-scaling sequence, and hence of the limiting diffusion? That is, does a different scaling also produce a (non-trivial) limit, or, rephrased, is the proposed scaling the correct one to look at? Secondly, the theorem deals with the norms for the range of $\alpha > d$ in the Hilbert scale, what can be said about convergence in the stronger norms corresponding to the range of $\alpha \in [0, d]$? Does there exist a limit? We conclude this section addressing these two issues.

Regarding the first question, it is immediately obvious that the re-scaling sequence $\frac{\ell_-(n)}{v_+(n)}$, which we denote by ρ_n in the following, is not a unique sequence yielding a non-trivial limit. Re-scaling the martingales M^n by any sequence of the form $\sqrt{c\rho_n}$ yields a convergent martingale sequence. However, the limiting diffusion differs only in a covariance operator, which is also re-scaled by c, and hence the limit is essentially the same process with either 'stretched' or 'shrinked' variability. However, the asymptotic behaviour of the re-scaling sequences, which allow for a non-trivial weak limit is unique. In general, by considering different re-scaling sequences ρ_n^* , we obtain three possibilities for the convergence of the sequence $\sqrt{\rho_n^*}M^n$. If ρ_n^* is of the same speed of convergence as ρ_n , i.e., for $\rho_n^* = \mathcal{O}(\rho_n)$, the thus re-scaled sequence converges again to a diffusion process for which the covariance operator is proportional to (2.18). This is then just a re-scaling by a sequence (asymptotically) proportional to ρ_n as discussed above. Secondly, if the convergence is slower, i.e., $\rho_n^* = o(\rho_n)$, then the same methods as in the law of large numbers show that the sequence converges to zero uniformly on compacts in probability, hence also convergence in distribution to the degenerate zero process follows. Thus, one only obtains the trivial limit. Finally, if we rescale by a sequence that diverges faster, i.e., $\rho_n = o(\rho_n^*)$, we can show that there does not exist a limit. This follows from general necessary conditions for the preservation of weak limits under transformation, which presuppose that $\sqrt{\rho_n^*/\rho_n M}$ has to converge in distribution in order for $\sqrt{\rho_n^* M_n}$ possessing a limit in distribution; see Theorem 2 in [29]. As the sequence ρ_n^*/ρ_n diverges, this is clearly not possible to hold.

Unfortunately, an answer to the second question is not possible in this clarity, when considering non-trivial limits. Essentially, we can only say that the currently used methods do not allow for any conclusion on convergence. The limitations are the following: The central problem is that for the parameter range $\alpha \in [0, d]$ the current method does not provide tightness of the re-scaled martingale sequence, hence we cannot infer that the sequence possesses a convergent subsequence. However, if tightness can be established in a different way then for the range $\alpha \in (\max\{1, d/2\}, d]$, the limit has to be the diffusion process defined by the operator (2.18) as follows from the characterisation of any limit in the proof of the theorem. Here, the lower bound of $\max\{1, d/2\}$ results, on the one hand, from our estimation technique, which necessitates α≥1 α≥1$\alpha \geq 1$, and on the other hand, from the definition of the limiting diffusion. Recall that the covariance operator is only of trace class for $\alpha > d/2$. Hence, $\alpha \in [0, d/2]$, we can no longer infer that the limiting diffusion even exists.

2.3 The Mean-Field Langevin Equation

An important property of the limiting diffusion in view toward analytic and numerical studies is that it can be represented by a stochastic integral with respect to a cylindrical or Q-Wiener process. For a general discussion of infinite-dimensional stochastic integrals, we refer to [12]. First, let $(W_t)_{t\geq 0}$ be a cylindrical Wiener process on $H^{-\alpha}(D)$ with covariance operator being the identity. Then $G(\nu(t), t) \circ \iota^{-1}$ is a trace class operator on $H^{-\alpha}(D)$ for suitable values of α. Here, $\iota^{-1} : H^{-\alpha}(D) \to H^{\alpha}(D)$ is the Riesz representation, i.e., the usual identification of a Hilbert space with its dual. The operator $G(\nu(t), t) \circ \iota^{-1}$ possesses a unique square-root we denote by $\sqrt{G(\nu(t), t) \circ \iota^{-1}}$, which is a Hilbert–Schmidt operator on $H^{-\alpha}(D)$. It follows that the stochastic integral process

$$Z_t := \int_0^t \sqrt{G(\nu(s), s) \circ \iota^{-1}}\, \mathrm{d}W_s \tag{2.20}$$

is a diffusion process in $H^{-\alpha}(D)$ with covariance operator $C(t)$. That is, $(Z_t)_{t\geq 0}$ is a version of the limiting diffusion in Theorem 2.3. Now, formally substituting for the limits in (2.15) yields the linear noise approximation

$$U_t = \nu_0 + \int_0^t \tau^{-1}\big(U_s + F(U_s, s)\big)\, \mathrm{d}s + \epsilon_n \int_0^t \sqrt{G(\nu(s), s) \circ \iota^{-1}}\, \mathrm{d}W_s.$$

or in differential notation

$$\mathrm{d}U_t = \tau^{-1}\big(U_t + F(U_t, t)\big)\,\mathrm{d}t + \epsilon_n\sqrt{G\big(v(t), t\big) \circ \iota^{-1}}\,\mathrm{d}W_t, \quad U_0 = v_0, \qquad (2.21)$$

where $\epsilon_n = \sqrt{v_+(n)/\ell_-(n)}$ is small for large n. Here, we have used the operator notation

$$F : H^{-\alpha}(D) \times \mathbb{R}_+ \to H^{-\alpha}(D) : F(g, t)(x) = f\big(\langle g, w(x, \cdot)\rangle_{H^\alpha(D)} + I(t, x)\big).$$

Equation (2.21) is an infinite-dimensional stochastic differential equation with additive (linear) noise. Here, additive means that the coefficient in the diffusion term does not depend on the solution U_t . A second formal substitution yields the Langevin approximation. Here, the dependence of the diffusion coefficient on the deterministic limit v is formally substituted by a dependence on the solution. That is, we obtain a stochastic partial differential equation with multiplicative noise given by

$$V_t = V_0 + \int_0^t \tau^{-1}\big(V_s + F(V_s, s)\big)\,\mathrm{d}s + \epsilon_n \int_0^t \sqrt{G(V_s, s) \circ \iota^{-1}}\,\mathrm{d}W_s,$$

or in differential notation

$$\mathrm{d}V_t = \tau^{-1}\big(V_t + F(V_t, t)\big)\,\mathrm{d}t + \epsilon_n\sqrt{G(V_t, t) \circ \iota^{-1}}\,\mathrm{d}W_t. \qquad (2.22)$$

Note that the derivation of the above equations was only formal, hence we have to address the existence and uniqueness of solutions and the proper setting for these equations. This is left for future work. It is an ongoing discussion and probably undecidable as lacking a criterion of approximation quality which—if any at all—is the correct diffusion approximation to use. First of all note that for both versions the noise term vanishes for $n \to \infty$, and thus both have the Wilson–Cowan equation as their limit. And also, neither of them approximates even the first moment of the microscopic models exactly. This means that for neither we have that the mean solves the Wilson–Cowan equation, which would be only the case if f were linear. However, they are close to the mean of the discrete process. We discuss this aspect in Appendix B.

Furthermore, we already observe in the central limit theorem, and thus also in the linear noise and Langevin approximation that the covariance (2.18) or the drift and the structure of the diffusion terms in (2.21) and (2.22), respectively, are independent of objects resulting from the microscopic models. They are defined purely in terms of the macroscopic limit. This observation supports the conjecture that these approximations are independent from possible different microscopic models converging to the same deterministic limit. Analogous statements hold also for derivations from the van Kampen system size expansion [5] and in related limit theorems for reaction diffusion models [4, 17,

18]. The only object reminiscent of the microscopic models in the continuous approximations is the re-scaling sequence ϵ_n . However, the re-scaling is proportional to the square root of $\ell_-(n)/v_+(n)$, i.e., the number of neurons per area divided by the size of the area, which is just the local density of particles. Therefore, in the approximations, the noise scales inversely to the square root of neuron density in this model, which interpreted in this way can also be considered a macroscopic fixed parameter and chosen independently of the approximating sequence.

Remark 2.4 The stochastic partial differential equations (2.21) and (2.22), which we proposed as the linear noise or Langevin approximation, respectively, are not necessarily unique as the representation of the limiting diffusion as a stochastic integral process (2.20) may not be unique. It will be subject for further research efforts to analyse the practical implications and usability of this Langevin approximation. Let Q be a trace class operator, $(W_t^Q)_{t\geq 0}$ be a Q-Wiener process and let B(v(t),t) B(v(t),t)$B(v(t),t)$ be operators such that $B(v(t),t)\circ Q\circ B(v(t),t)^* = G(v(t),t)\circ\iota^{-1}$, where $*$ denotes the adjoint operator. Then also the stochastic integral process

$$Z_t^Q := \int_0^t B\big(v(s),s\big)\,\mathrm{d}W_s^Q$$

is a version of the limiting diffusion in (2.3) and the corresponding linear noise and Langevin approximations are given by

$$\mathrm{d}U_t^Q = \tau^{-1}\big(U_t^Q + F(U_t^Q,t)\big)\,\mathrm{d}t + \epsilon_n B\big(v(t),t\big)\,\mathrm{d}W_t^Q$$

and

$$\mathrm{d}V_t^Q = \tau^{-1}\big(V_t^Q + F(V_t^Q,t)\big)\,\mathrm{d}t + \epsilon_n B\big(V_t^Q,t\big)\,\mathrm{d}W_t^Q.$$

We conclude this section by presenting one particular choice of a diffusion coefficient and a Wiener process. We take $(W_t^Q)_{t\geq 0}$ to be a cylindrical Wiener process on $L^2(D)$ with covariance $Q=\mathrm{Id}_{L^2}$. Then we can choose $B(t) = j\circ(\cdot\sqrt{g(t)})\in L(L^2(D),H^{-\alpha}(D))$, where j is the embedding operator $L^2(D)\hookrightarrow H^{-\alpha}(D)$ in the sense of (1.2) and $(\cdot\sqrt{g(t)})\in L(L^2(D),L^2(D))$ denotes a pointwise product of a function in $L^2(D)$, i.e.,

$$\big(\phi\cdot\sqrt{g(t)}\big)(x) = \phi(x)\bigg(\tau^{-1}v(t,x)+\tau^{-1}f\bigg(\int_D w(x,y)v(t,y)\,\mathrm{d}y + I(t,x)\bigg)\bigg)^{1/2}.$$

We first investigate the operator $G(v(t),t)\circ\iota^{-1}$ and write it in more detail as the following composition of operators:

$$G\big(v(t),t\big)\circ\iota^{-1} = j\circ\big(\cdot g(t)\big)\circ k\circ\iota^{-1},$$

where k is the embedding operator $H^{\alpha}(D) \hookrightarrow L^2(D)$. Next, the Hilbert adjoint $B^* \in L(H^{-\alpha}, L^2)$ is given by $B^* = (\cdot\sqrt{g}) \circ k \circ \iota^{-1}$, which is easy to verify. Hence, the stochastic integral of $B(t)$ with respect to W^Q is again a version of the limiting martingale as

$$\begin{aligned} B(t) \circ Q \circ B^*(t) &= j \circ \left(\cdot\sqrt{g(t)}\right) \circ \mathrm{Id}_{L^2} \circ \left(\cdot\sqrt{g(t)}\right) \circ k \circ \iota^{-1} \\ &= j \circ \left(\cdot g(t)\right) \circ k \circ \iota^{-1} = G\left(\nu(t), t\right) \circ \iota^{-1}. \end{aligned}$$

3 DISCUSSION AND EXTENSIONS

In this article, we have presented limit theorems that connect finite, discrete microscopic models of neural activity to the Wilson–Cowan neural field equation. The results state qualitative connections between the models formulated as precise probabilistic convergence concepts. Thus, the results strengthen the connection derived in a heuristic way from the van Kampen system size expansion.

A general limitation of mathematically precise approaches to approximations, cf. also the propagation to chaos limit theorems in [30], is that the microscopic models are usually defined via the limit. In other words, the limit has to be known a priori, and we look for models which converge to this limit. Thus, in contrast to the van Kampen system size expansion, the presented results are not a step-by-step modelling procedure in the sense that, via a constructive limiting procedure, a microscopic model yields a deterministic or stochastic approximation. Hence, it might be objected that the presented method can only be used a posteriori in order to justify a macroscopic model from a constructed microscopic model and that somehow one has to 'guess' the correct limit in advance. Several remarks can be made to answer this objection.

First, this observation is certainly true, but not necessarily a drawback. On the contrary, when both microscopic and macroscopic models are available, then it is rather important to know how these are connected and qualitatively and quantitatively characterise this connection. Concerning neural field models, this precise connection was simply not available so far for the well-established Wilson–Cowan model. Furthermore, when starting from a stochastic microscopic description working through proving the conditions for convergence for given microscopic models, one obtains very strong hints on the structure of a possible deterministic limit. Therefore, our results can also ease the procedure of 'guessing the correct limit'.

Secondly, often a phenomenological, deterministic model, which is an approximation to an inherently probabilistic process is derived from ad-hoc heuristic arguments. Given that the model has proved useful, one often aims to derive a justification from first principles and/or a stochastic version, which keeps the features of the deterministic model, but also accounts for the formerly neglected fluctuations. A standard, though somewhat simple approach to obtain stochastic versions consists of adding (small) noise to the deterministic equations. This article, provides a second approach which consists of finding microscopic models, which converge to the deterministic limit to obtain a stochastic correction via a central limit argument.

Thirdly and finally, the method also provides an argument for new equations, i.e., the Langevin and linear noise approximations, which can be used to study the stochastic fluctuations in the model. Furthermore, in contrast to previous studies, we do not provide deterministic moment equations but stochastic processes, which can be, e.g., via Monte Carlo simulations, studied concerning a large number of pathwise properties and dynamics beyond first and second moments.

We now conclude this article commenting on the feasibility of our approach connecting microscopic Markov models to deterministic macroscopic equations when dealing with different master equation formulations that appear in the literature. Additionally, the following discussions also relate the model (1.6) considered in this article to other master equation formulations. We conjecture that the analogous results as presented for the Wilson–Cowan equation (1.3) in Sect. 2 also hold for these variations of the master equations. This should be possible to achieve by an adaptation of the methods of proof presented although we have not performed the computations in detail.

3.1 A Variation of the Master Equation Formulation

A first variation of the discrete model we discussed in Sect. 1.2 was considered in the articles [8, 9] and a version restricted to a bounded state space also appears in [31]. This model consists of the master equation stated below in (3.2), which closely resembles (1.6). In the earlier reference [8], the model was introduced with a different interpretation called the effective spike model. We briefly explain this interpretation before presenting the master equation. Instead of interpreting P as the number of neuron populations, in this model, P denotes the number of different neurons in the network located within a spatial domain D. Then Θ_t^k, the state of the k th neuron, counts the number of 'effective' spikes this neuron has emitted in the past up till time t. Effective spikes are those spikes that still influence the dynamics of the system, e.g., via a post-synaptic potential. Then state transitions adding/subtracting one effective spike for the k th neuron are governed by a firing rate function $\tilde{f}_k$, which depends on the input into neuron k, and a decay rate τ^{-1}. The constant decay rate indicates that emitted spikes are effective for a time interval of length τ and the gain function is defined—neglecting external input—by

$$\widetilde{f}_k(\theta) = f^*\left(\sum_{j=1}^{P} \widetilde{W}_{kj}\theta^j\right).$$

where f^* is a certain non-negative, real function. It is stated clearly in [9] that the function f^* is not equal to the gain function f in the proposed limiting Wilson–Cowan equation (1.3), but rather connected to f such that

$$\mathbb{E}f^*\left(\sum_{j=1}^{P} \widetilde{W}_{kj}\Theta_t^j\right) = f\left(\sum_{j=1}^{P} \widetilde{W}_{kj}\mathbb{E}\Theta_t^j\right) + \text{higher order terms.} \tag{3.1}$$

The authors in [9] state that for any function f such a function f^* can be found. Then the process $\Theta_t = (\Theta_t^1, \ldots, \Theta_t^P)$ is a jump Markov process given by the master equation

$$\frac{\mathrm{d}\mathbb{P}[\theta, t]}{\mathrm{d}t} = \sum_{k=1}^{P}\Bigg[\widetilde{f}_k(\theta - e_k)\mathbb{P}[\theta - e_k, t] - \left(\frac{1}{\tau}\theta^k + \widetilde{f}_k(\theta)\right)\mathbb{P}[\theta, t] + \frac{1}{\tau}(\theta^k + 1)\mathbb{P}[\theta + e_k, t]\Bigg] \tag{3.2}$$

with boundary conditions $\mathbb{P}[\theta, t] = 0$ if $\theta \notin \mathbb{N}_0^P$ as stated in [9]. The advantage of the effective spike model interpretation over the interpretation as neurons per population is that the unbounded state space of the model is justified. In principle, there can be an arbitrary number of spikes emitted in the past still active. However, a disadvantage of the master equation (3.2) is that for taking the limit it lacks a parameter corresponding to the system size providing a natural small parameter in the van Kampen system size expansion. This explains the shift in the interpretation of the master equation in the study [9] following [8], and subsequently in [5] to the interpretation we presented in Sect. 1.2, which provides the system-size parameters $l(k)$.

On the level of Markov jump processes, the master equation (3.2) obviously describes dynamics similar to the master equation (1.6) only replacing the activation rate $\tau^{-1}l(k)\overline{f}_k(\theta)$ in (1.6) by $\widetilde{f}_k(\theta)$ which is independent of the parameter $l(k)$. Thus, the model (3.2) can be understood as resulting from (1.6) after a limit procedure taking $l(k) \to \infty$ has been applied and the firing rate functions are connected via the forma $\lim_{l(k)\to\infty} l(k)\overline{f}_k(\theta) = \widetilde{f}_k(\theta)$. A qualitative interpretation of this limit procedure connecting the two types of models is given in [8]. This observation motivated the model in [5] stepping back one limit procedure, and thus providing the correct framework for the derivation of limit theorems.

It would be an interesting addition to the limit theorems in Theorem 2.1 to derive a law of large numbers for the models (3.2) with stochastic mean activity ν^n as defined in (2.7) and suitable chosen weights $\widetilde{W}_{kj}$. Clearly, the macroscopic limit should be given by the Wilson–Cowan equation (1.3). We conjecture that the appropriate condition for the function f^* in the present setting—including time dependent inputs—is

$$\mathbb{E}\left[l(k,n)^{-1} f^*\left(\sum_{j=1}^{P} \widetilde{W}^n_{kj}\Theta^j + \widetilde{I}_{k,n}(t)\right)\right]$$

$$= f\left(\sum_{j=1}^{P} \overline{W}^n_{kj}\frac{\mathbb{E}\Theta_j}{l(j,n)} + \overline{I}_{k,n}(t)\right) + \text{h.o.t.}, \tag{3.3}$$

such that the higher order terms are uniformly bounded and vanish in the limit $n \to \infty$, and where the weights $\overline{W}^n_{kj}$ and inputs $\overline{I}_{k,n}(t)$ are defined as in (2.4) and (2.6). Property (3.3) closely resembles condition (3.1) and trivially holds for linear f with $f^* = f$.

3.2 Bounded State Space Master Equations

We have already stated when introducing the microscopic model in Sect. 1.2 that the interpretation of the parameter $l(k)$ as the number of neurons in the k th population is not literally correct. The state space of the process is unbounded, hence arbitrarily many neurons can be active, and thus each population contains arbitrarily many neurons. In order to overcome this interpretation problem, it was supposed to consider the master equation only on a bounded state space. That is, the k th population consists of $l(k)$ neurons, and $0 \le \Theta^k_t \le l(k)$ almost surely. Such master equations are simply obtained by setting the transition rates for transition of θ^k from $l(k) \to l(k)+1$ to zero.

A first master equation of this form was considered in [22], which in present notation, takes the form

$$\frac{\mathrm{d}\mathbb{P}[\theta,t]}{\mathrm{d}t} = \frac{1}{\tau}\sum_{k=1}^{P}\big[\big(l(k)-\theta^k+1\big)\overline{f}_k(\theta-e_k)\mathbb{P}\big[\theta^k-e_k,t\big]$$
$$-\big(\theta^k+\big(l(k)-\theta^k\big)\overline{f}_k(\theta)\big)\mathbb{P}[\theta,t]+\big(\theta^k+1\big)\mathbb{P}[\theta+e_k,t]\big]. \tag{3.4}$$

Versions of such a master equation for, e.g., one population only or coupled inhibitory and excitatory populations were considered in [3, 22], and a van Kampen systems size expansion was carried out. Here, the bound in the state space provides a natural parameter for the re-scaling, thus a small parameter for the expansion. The setup of this problem resembles closely the structure of excitable membranes for which limits have been obtained with the present technique by one of the present author and co-workers in [27]. Therefore, we conjecture that our limit theorems also apply to this setting with minor adaptations with essentially the same conditions and results as in Sect. 2. However, the macroscopic limit, which will be obtained does not conform with the Wilson–Cowan equation but will be given by

$$\tau \dot{\nu}(t,x) = -\nu(t,x) + \big(1-\nu(t,x)\big) f\left(\int_D w(x,y)\nu(t,y)\,\mathrm{d}y + I(t,x) \right). \tag{3.5}$$

Next, we return to the master equation (1.6) as discussed in this article in Sect. 1.2 and the comment we made regarding bounded state spaces the footnote on page 7. In our primary reference for this model [5], actually a bounded state space version of the master equation was considered where the activation rate for the event $\theta^k \to \theta^k + 1$ is

$$l(k)\overline{f}_k(\theta,t)\mathbb{I}_{[\theta_k < l(k)]}, \tag{3.6}$$

replacing $l(k)\overline{f}_k(\theta,t)$ in (1.6). The van Kampen system size expansion was then applied to this bounded state space master equation, tacitly neglecting possible difficulties, which might arise due to the discontinuity of (3.6) considered as a function on $\mathbb{R}^P$. However, for the present, mathematically precise limit convergence results considering bounded state space as originally suggested in [5] are problematic. The discontinuous activation rate (3.6) causes the machinery developed in [27], which depends on Lipschitz-type estimates to break down. However, we strongly expect that also in this case the law of large numbers with the deterministic limit given by the Wilson–Cowan equation (1.3) holds. Furthermore, also the Langevin approximations should agree with the equations discussed in Sect. 2.3. However, we have not yet been able to prove such a theorem. We further conjecture that the results in this article can be used to prove the convergence for the bounded state space model by a domination argument. Heuristically, it seems clear that a bounded process should be dominated by a process that possesses the same dynamics inside the state space of the bounded process, but can stray out from that bounded domain. Hence, as the limit of the potentially larger process lies within the domain where the two processes agree also the dominated process should converge to the same limit. Mathematically, this line of argument relies on non-trivial estimates between occupation measures of high-dimensional Markov processes. This is work in progress.

3.3 Activity Based Neural Field Model

Finally, we return also to a difference in neural field theory mentioned in the beginning. In contrast to rate-based neural field models of the Wilson–Cowan type (1.1), there exists a second essential class of neural field models, so-called activity based models, the prototype of which is the Amari equation

$$\tau \dot{v}(t,x) = -v(t,x) + \int_D w(x,y) f\big(v(t,y)\big)\,\mathrm{d}y + I(t,x). \tag{3.7}$$

We conjecture that also for this type of equations a phenomenological microscopic model can be constructed with a suitable adaptation of the activation rates and that limit theorems analogous to the results in Sect. 2.1 hold. Then also a Langevin equation for this model can be obtained and used for further analysis.

4 PROOFS OF THE MAIN RESULTS

In this section, we present the proofs of the limit theorems. For the convenience of the reader, as it is important tool in the subsequent proofs, we first state the Poincaré inequality. Let $D \subset \mathbb{R}^d$ be a convex domain, then it holds for any function $\phi \in H^1(D)$ that

$$\|\overline{\phi}_D - \phi\|_{L^2(D)} \le \frac{\mathrm{diam}(D)}{\pi} \|\nabla\phi\|_{L^2(D)}. \tag{4.1}$$

where $\overline{\phi}_D$ is the mean value of the function ϕ on the domain D, i.e.,

$$\overline{\phi}_D = \frac{1}{|D|} \int_D \phi(x)\,\mathrm{d}x. \tag{4.2}$$

Moreover, the constant in the right-hand side of (4.1) is the optimal constant depending only on the diameter of the domain D, cf. [1, 23]. Whenever we omit to denote the spatial domain for definition of norms or inner products in $L^2(D)$ or Sobolev spaces $H^\alpha(D)$, then it is to be interpreted as the norm over the whole domain D. If the norm is taken only over a subset $D_{k,n}$, then this is always indicated unexceptionally.

$$\tau \dot{v}(t,x) = -v(t,x) + f\left(\int_D w(x,y) v(t,y)\,\mathrm{d}y + I(t,x)\right). \tag{4.3}$$

We denote by F the Nemytzkii operator on L 2 (D) L2(D)$L^2(D)$ defined by

$$F(g,t)(x) = f\left(\int_D w(x,y)g(y)\,\mathrm{d}y + I(t,x)\right) \quad \forall g \in L^2(D), \tag{4.4}$$

and for all $\theta \in \mathbb{N}_0^P$ we define a discrete version of the Nemytzkii operator via

$$\begin{aligned} -\frac{1}{\tau}v^n(\theta) + \frac{1}{\tau}\overline{F}^n\big(v^n(\theta),t\big) &= \lambda^n(\theta,t)\int_{\mathbb{N}_0^P}\big(v^n(\xi) - v^n(\theta)\big)\mu^n\big((\theta,t),\mathrm{d}\xi\big) \\ &= \frac{1}{\tau}\sum_{k=1}^{P}\frac{1}{l(k,n)}\big(-\theta^k + l(k,n)\overline{f}_{k,n}(\theta,t)\big)\mathbb{I}_{D_{k,n}} \\ &= -\frac{1}{\tau}v^n(\theta) + \frac{1}{\tau}\sum_{k=1}^{P}\overline{f}_{k,n}(\theta,t)\mathbb{I}_{D_{k,n}}. \end{aligned} \tag{4.5}$$

Note that $\tau^{-1}(\phi, v^n(\theta))_{L^2} + \tau^{-1}(\phi, \overline{F}^n(v^n(\theta),t))_{L^2}$ for $\phi \in L^2(D)$ corresponds to the generator of $(\Theta_t^n, t)_{t\geq 0}$ applied to the function $(\theta,t) \mapsto (\phi, v^n(\theta))_{L^2}$.

Finally, another useful property is that the means of the process' components are bounded. For each k, n it holds that

$$\begin{aligned} \mathbb{E}\Theta_t^{k,n} &= \mathbb{E}\Theta_0^{k,n} + \frac{1}{\tau}\int_0^t l(k,n)\mathbb{E}\overline{f}_{k,n}(Y_s^n) - \mathbb{E}\Theta_s^{k,n}\,\mathrm{d}s \\ &\leq \mathbb{E}\Theta_0^{k,n} + \frac{1}{\tau}\int_0^t l(k,n)\|f\|_0 - \mathbb{E}\Theta_s^{k,n}\,\mathrm{d}s, \end{aligned}$$

see also (B.1). Therefore, it holds that $\mathbb{E}\Theta_t^{k,n} \leq m_t^{k,n}$, where $m_t^{k,n}$ solves the deterministic initial value problem

$$\dot{m}_t^{k,n} = -\frac{1}{\tau}m_t^{k,n} + \frac{1}{\tau}l(k,n)\|f\|_0, \quad m_0^{k,n} = \mathbb{E}\Theta_0^{k,n}.$$

i.e.,

$$\begin{aligned} m_t^{k,n} &= \mathrm{e}^{-t/\tau}\big(m_{k,n}^0 - l(k,n)\|f\|_0\big) + l(k,n)\|f\|_0 \\ &\leq l(k,n)\big(1 + \|f\|_0\big) \quad \forall t \geq 0. \end{aligned} \tag{4.6}$$

Here, we also used the assumption $\mathbb{E}^n\Theta_0^{k,n} \leq l(k,n)$ on the initial condition.

4.1 Proof of Theorem 2.1 (Law of Large Numbers)

In order to prove the law of large numbers, Theorem 2.1, we apply the law of large numbers for Hilbert space valued PDMPs, see Theorem 4.1 in [27], to the sequence of homogeneous PDMPs $(Y_t^n)_{t\geq 0} = (\Theta_t^n, t)_{t\geq 0}$. For the application of this theorem, recall that the first, piecewise constant, vector-valued component of this process counts the number of active neurons in each sub-population and the second, deterministic component states time. The process $(Y_t^n)_{t\geq 0}$ is the usual 'space-time process', i.e., homogeneous Markov process which is obtained via a state-space extension to obtain a homogeneous Markov process from the inhomogeneous process $(\Theta_t^n)_{t\geq 0}$. The continuous component satisfies the simple ODE $\dot{t} = 1$, $t(0) = 0$, and thus the full process is a PDMP. In the terminology of [27], the sequence of coordinate functions on the different state spaces of the PDMPs $(Y_t^n)_{t\geq 0}$ into a common Hilbert space is given by the maps ν^n (2.7) with the common Hilbert space $L^2(D)$. Thus, in order to infer convergence in probability (2.10) from Theorem 4.1 in [27], it is sufficient to validate the following conditions:

(LLN1) For fixed $T > 0$, it holds that

$$\lim_{n\to\infty} \mathbb{E}^n \int_0^T \lambda^n(Y_t^n) \int_{\mathbb{N}^P} \left\| \nu^n(\xi) - \nu^n(\Theta_t^n) \right\|_{L^2}^2 \mu^n(Y_t^n, \mathrm{d}\xi)\, \mathrm{d}t = 0. \tag{4.7}$$

(LLN2) The Nemytzkii operator F satisfies a Lipschitz condition in $L^2(D)$ uniformly with respect to t, t≥0 t≥0$t \geq 0$, i.e., there exists a constant L 0 >0 L0>0$L_0 > 0$ such that

$$\left\| F(g_1, t) - F(g_2, t) \right\|_{L^2} \leq L_0 \|g_1 - g_2\|_{L^2} \quad \forall t \geq 0, g_1, g_2 \in L^2(D). \tag{4.8}$$

(LLN3) For fixet $T > 0$, it holds that

$$\lim_{n\to\infty} \mathbb{E}^n \int_0^T \left\| \overline{F}^n(\nu_t^n, t) - F(\nu_t^n, t) \right\|_{L^2} \mathrm{d}t = 0. \tag{4.9}$$

Note that the final condition of Theorem 4.1 in [27], i.e., the convergence of the initial conditions, is satisfied by assumption. For a discussion of these conditions, we refer to [27] and proceed to their derivation for the present model in the subsequent parts (a) to (c).

(a)In order to prove condition (4.7), we write the integral with respect to the discrete probability measure μ^n as a sum. This yields

$$\begin{aligned}
&\mathbb{E}^n\lambda(Y_t^n)\int_{\mathbb{N}^P}\|\nu^n(\xi)-\nu^n(\Theta_t^n)\|_{L^2}^2\mu^n(Y_t^n,\mathrm{d}\xi)\\
&=\frac{1}{\tau}\sum_{k=1}^{P}\mathbb{E}^n\frac{1}{l(k,n)^2}(\Theta_t^{k,n}+l(k,n)\overline{f}_{k,n}(Y_t^n))|D_{k,n}|\\
&\leq\frac{1}{\tau}\frac{1+2\|f\|_0}{\ell_-(n)}|D|.
\end{aligned}\tag{4.10}$$

where we have used the upper bound (4.6) on the expectation $\mathbb{E}^n\Theta_t^{k,n}$ and the assumption on the initial conditions. Next, integrating over [0,T] [0,T]$[0,T]$ and employing the assumption $\lim_{n\to\infty}\ell_-(n)=\infty$ in (2.9) establishes condition (4.7).

(b)The Lipschitz condition (4.8) of the Nemytzkii operators is a straightforward consequence of the Lipschitz continuity (1.4) of the gain function f as

$$\begin{aligned}
&\|F(g_1,t)-F(g_2,t)\|_{L^2}^2\\
&=\int_D\left|f\left(\int_D w(x,y)g_1(y)\,\mathrm{d}y+I(x,t)\right)\right.\\
&\qquad\left.-f\left(\int_D w(x,y)g_2(y)\,\mathrm{d}y+I(x,t)\right)\right|^2\mathrm{d}x\\
&\leq L^2\int_D\left|\int_D w(x,y)(g_1(y)-g_2(y))\,\mathrm{d}y\right|^2\mathrm{d}x\\
&\leq L^2\int_D\|w(x,\cdot)\|_{L^2}^2\|g_1-g_2\|_{L^2}^2\,\mathrm{d}x\\
&=L^2\|w\|_{L^2\times L^2}^2\|g_1-g_2\|_{L^2}^2.
\end{aligned}$$

Therefore, (4.8) holds with Lipschitz constant $L_0:=L\|w\|_{L^2\times L^2}$.

(c)Finally, we prove the convergence of the generators (4.9). To this end, we employ the characterisation of the norm in $L^2(D)$ by $|\|\eta\|_{L^2}=\sup_{\|\phi\|_{L^2}=1}|(\phi,\eta)_{L^2}|$ for all ηEL 2 (D))$\eta\in L^2(D)$, and thus consider first the scalar product of elements $\phi\in L^2(D)$ with $\|\phi\|_{L^2}=1$ and the difference

$$(\phi,\overline{F}^n(\nu_t^n,t))_{L^2}=\left(\phi,\sum_{k=1}^{P}\overline{f}_{k,n}(Y_t^n)\mathbb{I}_{D_{k,n}}\right)_{L^2}.\tag{4.11}$$

Next, we apply the Nemytzkii operator F defined in (4.4) to v n (t) vn(t)$v^n(t)$ and take the inner product of the result with respect to ϕ to obtain on the other hand

$$(\phi, F(v_t^n, t))_{L^2} = \left(\phi, f\left(\sum_{k=1}^{P} \frac{\Theta_t^{k,n}}{l(k,n)} \int_{D_{k,n}} w(\cdot, y)\,\mathrm{d}y + I(t, \cdot)\right)\right)_{L^2} \tag{4.12}$$

Subtracting (4.12) from (4.11), we obtain the integrated difference

$$\begin{aligned}
&(\phi, \overline{F}^n(v_t^n, t))_{L^2} - (\phi, F(v_t^n, t))_{L^2} \\
&= \int_D \phi(x)\left[\sum_{k=1}^{P} \overline{f}_{k,n}(Y_t^n)\mathbb{I}_{D_{k,n}}\right. \\
&\quad \left. - f\left(\sum_{j=1}^{P} \frac{\Theta_j^n(t)}{l(j,n)} \int_{D_{j,n}} w(x, y)\,\mathrm{d}y + I(t, x)\right)\right]\mathrm{d}x \\
&= \sum_{k=1}^{P} \int_{D_{k,n}} \phi(x)\left[f\left(\sum_{j=1}^{P} \overline{W}_{kj}^n \frac{\Theta_t^{j,n}}{l(j,n)} + \overline{I}_{k,n}(t)\right)\right. \\
&\quad \left. - f\left(\sum_{j=1}^{P} \frac{\Theta_t^{j,n}}{l(j,n)} \int_{D_{j,n}} w(x, y)\,\mathrm{d}y + I(t, x)\right)\right]\mathrm{d}x.
\end{aligned}$$

We proceed to estimate the norm of the term in the right-hand side. We use the Lipschitz condition (1.4) on f, the triangle inequality, and finally the Cauchy–Schwarz inequality on the resulting second term to obtain the estimate

$$\begin{aligned}
&\left|(\phi, \overline{F}^n(v_t^n, t))_{L^2} - (\phi, F(v_t^n, t))_{L^2}\right| \\
&\leq L \sum_{k=1}^{P} \int_{D_{k,n}} |\phi(x)| \\
&\quad \times \left|\sum_{i=1}^{P} \frac{\Theta_j^n(t)}{l(j,n)}\left(\overline{W}_{kj}^n - \int_{D_{j,n}} w(x, y)\,\mathrm{d}y\right) + \overline{I}_{k,n}(t) - I(t, x)\right| \mathrm{d}x \\
&\leq \underbrace{L \sum_{k=1}^{P} \int_{D_{k,n}} |\phi(x)| \left|\sum_{j=1}^{P} \frac{\Theta_j^n(t)}{l(j,n)}\left(\overline{W}_{kj}^n - \int_{D_{j,n}} w(x, y)\,\mathrm{d}y\right)\right| \mathrm{d}x}_{(*)} \\
&\quad + \underbrace{L \sum_{k=1}^{P} \|\phi\|_{L^2(D_{k,n})} \left\|\overline{I}_{k,n}(t) - I(t)\right\|_{L^2(D_{k,n})}}_{(**)}.
\end{aligned}$$

Here, the term in the right-hand side marked $(**)$ is further estimated using the Cauchy–Schwarz inequality and the Poincaré inequality (4.1), which yields

$$(**) \leq \frac{\delta_+(n)}{\pi}\left(\sum_{k=1}^{P}\left\|\nabla I(t)\right\|^2_{L^2(D_{k,n})}\right)^{1/2} = \frac{\delta_+(n)}{\pi}\left\|\nabla_x I(t)\right\|_{L^2}. \tag{4.13}$$

We now consider the term marked $(*)$. Inserting the definition of $\overline{W}^n_{kj}$ given in (2.4), the reordering of the summations and changing the order of integration yields

$$\begin{aligned}(*) &= \sum_{k=1}^{P}\int_{D_{k,n}}\left|\phi(x)\right| \\ &\quad\times\left|\sum_{j=1}^{P}\frac{\Theta_t^{j,n}}{l(k,n)}\left(\int_{D_{j,n}}\left(\frac{1}{|D_{k,n}|}\int_{D_{k,n}} w(z,y)\,\mathrm{d}z\right) - w(x,y)\,\mathrm{d}y\right)\right|\mathrm{d}x \\ &\leq \sum_{k=1}^{P}\int_{D_{k,n}}\left|\phi(x)\right| \\ &\quad\times\sum_{j=1}^{P}\frac{\Theta_t^{j,n}}{l(k,n)}\int_{D_{j,n}}\left|\left(\frac{1}{|D_{k,n}|}\int_{D_{k,n}} w(z,y)\,\mathrm{d}z\right) - w(x,y)\right|\mathrm{d}y\,\mathrm{d}x \\ &\quad\times\sum_{j=1}^{P}\frac{\Theta_t^{j,n}}{l(k,n)}\int_{D_{j,n}}\left|\left(\frac{1}{|D_{k,n}|}\int_{D_{k,n}} w(z,y)\,\mathrm{d}z\right) - w(x,y)\right|\mathrm{d}y\,\mathrm{d}x \\ &= \sum_{k=1}^{P}\sum_{j=1}^{P}\frac{\Theta_t^{j,n}}{l(k,n)} \\ &\quad\times\int_{D_{j,n}}\left[\int_{D_{k,n}}\left|\phi(x)\right|\left|\left(\frac{1}{|D_{k,n}|}\int_{D_{k,n}} w(z,y)\,\mathrm{d}z\right) - w(x,y)\right|\mathrm{d}x\right]\mathrm{d}y.\end{aligned}$$

We next apply the Cauchy–Schwarz inequality to the integral inside the square brackets in the last term. Thus, we obtain the estimate

$$\begin{aligned}(*) &\leq \sum_{j=1}^{P}\int_{D_{j,n}}\sum_{k=1}^{P}\left\|\phi\right\|_{L^2(D_{k,n})} \\ &\quad\times\left[\int_{D_{k,n}}\left|\left(\frac{1}{|D_{k,n}|}\int_{D_{k,n}} w(z,y)\,\mathrm{d}z\right) - w(x,y)\right|^2\mathrm{d}x\right]^{1/2}\mathrm{d}y.\end{aligned}$$

Now the Poincaré inequality (4.1) is applied to the innermost integral inside the square brackets, which yields

$$(*) \leq \sum_{j=1}^{P} \frac{\Theta_t^{j,n}}{l(k,n)} \int_{D_{j,n}} \sum_{k=1}^{P} \|\phi\|_{L^2(D_{k,n})} \frac{\operatorname{diam}(D_{k,n})}{\pi} \left\| \nabla_x w(\cdot, y) \right\|_{L^2(D_{k,n})} \mathrm{d}y.$$

Finally, using once more the Cauchy–Schwarz inequality on the innermost summation we obtain

$$(*) \leq \frac{\delta_+(n)}{\pi} \sum_{j=1}^{P} \frac{\Theta_t^{j,n}}{l(k,n)} \int_{D_{j,n}} \left\| \nabla_x w(\cdot, y) \right\|_{L^2} \mathrm{d}y. \tag{4.14}$$

Now, a combination of the estimates (4.13) and (4.14) on the terms $(*)$ and $(**)$ yields

$$\begin{aligned} &\left| \left(\phi, \overline{F}^n(v_t^n, t)\right)_{L^2} - \left(\phi, F(v^n(t), t)\right)_{L^2} \right| \\ &\leq \delta_+(n) \frac{L}{\pi} \left(\sum_{j=1}^{P} \frac{\Theta_t^{j,n}}{l(k,n)} \int_{D_{j,n}} \left\| \nabla_x w(\cdot, y) \right\|_{L^2} \mathrm{d}y + \left\| \nabla_x I(t) \right\|_{L^2} \right). \end{aligned}$$

Here, the right-hand side is independent of ϕ, hence taking the supremum over all ϕ with $\|\phi\|_{L^2} = 1$ yields

$$\begin{aligned} &\left\| \overline{F}^n(v_t^n, t) - F(v^n(t), t) \right\|_{L^2} \\ &\leq \delta_+(n) \frac{L}{\pi} \left(\sum_{j=1}^{P} \frac{\Theta_t^{j,n}}{l(k,n)} \int_{D_{j,n}} \left\| \nabla_x w(\cdot, y) \right\|_{L^2} \mathrm{d}y + \left\| \nabla_x I(t) \right\|_{L^2} \right). \end{aligned}$$

Finally, integrating over $(0, T)$ and taking the expectation on both sides results in

$$\begin{aligned} &\mathbb{E}^n \int_0^T \left\| \overline{F}^n(v_t^n, t) - F(v^n(t), t) \right\|_{L^2} \mathrm{d}t \\ &\leq \delta_+(n) \frac{L}{\pi} \left(\sqrt{|D|} T (1 + \|f\|_0) \|\nabla_x w\|_{L^2 \times L^2} + \|\nabla_x I\|_{L^1((0,T),L^2)} \right). \end{aligned} \tag{4.15}$$

Here, we have used (4.6) and a combination of the Cauchy–Schwarz and Poincaré inequality (4.1) in order to estimate

$$\mathbb{E}^n \sum_{j=1}^{P} \frac{\Theta_t^{j,n}}{l(k,n)} \int_{D_{j,n}} \left\| \nabla_x w(\cdot, y) \right\|_{L^2} \mathrm{d}y \leq \delta_+(n) \frac{\sqrt{|D|}(1 + \|f\|_0)}{\pi} \|\nabla_x w\|_{L^2 \times L^2}.$$

The upper bound in (4.15) is of order $\mathcal{O}(\delta_+(n))$ and, therefore, converges to zero for $n \to \infty$ due to assumption (2.9). Hence, condition (4.9) is satisfied. The proof of the convergence in probability (2.10) is completed.

It is now easy to extend this result to the convergence in the r th mean. First of all, the convergence in probability (2.10) implies for all $r \geq 1$ the convergence in probability of the random variables $\sup_{t\in[0,T]} \|\nu_t^n - \nu(t)\|_{L^2}^r$ to zero. As convergence in the mean of real valued random variables is equivalent to convergence in probability and uniform integrability it remains to prove the latter for the families $\sup_{t\in[0,T]} \|\nu_t^n - \nu(t)\|_{L^2}^r$,$n \in \mathbb{N}$.

We first consider the case $r = 1$, and establish a uniform bound on the second moments $\mathbb{E}^n \sup_{t\in[0,T]} \|\nu_t^n - \nu(t)\|_{L^2}^2$. Then the de la Vallée–Poussin theorem, cf. App., Proposition 2.2 in [15], implies that the random variables $\sup_{t\in[0,T]} \|\nu_t^n - \nu(t)\|_{L^2}$, $n \in \mathbb{N}$, are uniformly integrable.

Without loss of generality, we can assume that there existe Poisson processes $(N_t^{k,n})_{t\geq 0}$ with rates $\Lambda_{k,n} = l(k,n)(1 + \|f\|_0)/\tau$, which dominate $(\Theta_t^{k,n} - \Theta_0^{k,n})_{t\geq 0}$ pathwise. Then we obtain almost surely

$$\|\nu_t^n\|_{L^2}^2 \leq 2\|\nu_0^n\|_{L^2}^2 + 2\sum_{k=1}^{P} \frac{(\Theta_t^{k,n} - \Theta_0^{k,n})^2}{l(k,n)^2}|D_{k,n}| \leq 2\|\nu_0^n\|_{L^2}^2 + 2\sum_{k=1}^{P} \frac{(N_T^{k,n})^2}{l(k,n)^2}|D_{k,n}|.$$

Here, the right-hand side is independent of t≤T t≤T$t \leq T$, and thus we obtain

$$\mathbb{E}^n \sup_{t\in[0,T]} \|\nu_t^n\|_{L^2}^2 \leq 2\mathbb{E}^n\|\nu_0^n\|_{L^2}^2 + 2\sum_{k=1}^{P} \frac{\mathbb{E}^n(N_T^{k,n})^2}{l(k,n)^2}|D_{k,n}| \leq 2\mathbb{E}^n\|\nu_0^n\|_{L^2}^2 + C_T,$$

where we have used that $N_T^{k,n}$ is Poisson distributed with rate $T\Lambda_{k,n}$, and thus $_2\mathbb{E}^n(N_T^{k,n})^2 = T\Lambda_{k,n} + T^2\Lambda_{k,n}^2$. Here, C_T is some finite constant which depends on T and the overall parameters of the model, i.e., τ, f, D, but is independent of k and n. Using this upper bound, the triangle inequality yields the estimate

$$\mathbb{E}^n \sup_{t\in[0,T]} \|\nu_t^n - \nu(t)\|_{L^2}^2 \leq 2C_T\big(\mathbb{E}^n\|\nu_0^n\|_{L^2}^2 + \|\nu\|_{C([0,T],L^2)}^2 + 1\big).$$

Therefore, using the assumption $\sup_{n\in\mathbb{N}} \mathbb{E}^n\|\nu_0^n\|_{L^2}^2 < \infty$ it holds that

$$\sup_{n\in\mathbb{N}} \mathbb{E}^n \sup_{t\in[0,T]} \|\nu_t^n - \nu(t)\|_{L^2}^2 < \infty.$$

The general case for $r > 1$ works analogously. Note that the r th moment of the Poisson distribution is proportional to the r th power of its rate. Hence, just as in the case of r=1 r=1$r = 1$, the term

$$\mathbb{E}^n \left(\sum_{k=1}^{P} \frac{(N_T^{k,n})^2}{l(k,n)^2} |D_{k,n}| \right)^r$$

can thus be bounded from above by some constant C_T independent of k and n. The proof of Theorem 2.1 is completed.

4.2 Proof of Corollary 2.1 (Corollary to the Law of Large Numbers)

For α=0 , the statement of the corollary coincides with the statement of Theorem 2.1, hence we consider α>0. As in the proof of Theorem 2.1, we apply Theorem 4.1 in [27] to the PDMPs $(Y_t^n)_{t\geq 0}$, however, this time for the functions $\mathrm{n}\nu^n$ understood as taking values in the Hilbert space $H^{-\alpha}(D)$ instead $L^2(D)$. Thus, we have to validate again conditions (LLN1)–(LLN3) wherein the norm in $L^2(D)$ is always replaced by the norm in $H^{-\alpha}(D)$. The essential argument is sharpening the estimates in part (a) of the proof of Theorem 2.1 using optimal Sobolev embedding theorems such that the conditions (2.13) imply (LLN1). This we present in part (a) of the proof below. The Lipschitz condition (LLN2) of the Nemytzkii operator F in the spaces $H^{-\alpha}$ is established in part (b). Finally, as the condition $\delta_+(n) \to \infty$ remains as in Theorem 2.1, the condition (LLN3) follows immediately from the proof of Theorem 2.1 due to the continuous embedding of $L^2(D)$ into $H^{-\alpha}(D)$.

(a)In the case α=0 , i.e., $H^\alpha(D) = L^2(D)$, we used in (4.10) that $\|\mathbb{I}_{D_{k,n}}\|_{L^2}^2 = |D_{k,n}|$. For general α>0, we use the representation

$$\|\mathbb{I}_{D_{k,n}}\|_{H^{-\alpha}} = \sup_{\|\phi\|_{H^\alpha}} \left|(\phi, \mathbb{I}_{D_{k,n}})_{L^2}\right|.$$

In order to estimate the terms inside the supremum in the right-hand side, we use Hölder's inequality and the Sobolev embedding theorem, i.e., $H^\alpha(D) \hookrightarrow L^\infty(D)$ for $\alpha > d/2$ and $H^\alpha(D) \hookrightarrow L^r(D)$ with $r = d/(d/2 - \alpha)$ for $0 < \alpha < d/2$, see Theorem 7.34 and Corollary 7.17 in [2]. Thus, we obtain

$$\|\mathbb{I}_{D_{k,n}}\|_{H^{-\alpha}} = \sup_{\|\phi\|_{H^\alpha}} \left|(\phi, \mathbb{I}_{D_{k,n}})_{L^2}\right|.$$

where the constants K are the constants arising from the continuous embeddings of the Sobolev spaces into the Lebesgue spaces. Evaluating the norms in the right-hand side, and further estimating using the maximal Lebesgue measure of the elements of the partition yields

$$\|\mathbb{I}_{D_{k,n}}\|_{H^{-\alpha}}^2 \le \begin{cases} K_{d/(d/2-\alpha)}^2 |D_{k,n}| v_+(n)^{2\alpha/d} & \text{if } 0 < \alpha < d/2, \\ K_\infty^2 |D_{k,n}| v_+(n) & \text{if } d/2 < \alpha. \end{cases}$$

Note that the upper bounds are consistent with the condition in Theorem 2.1 for $\alpha = 0$. Finally, as $H^{d/2}(D) \hookrightarrow H^{(d/2-\epsilon)}(D)$ for all small ϵ, the result for α=d/2 follows from the result above as

$$\|\mathbb{I}_{D_{k,n}}\|_{H^{-d/2}} \le \sup_{\|\phi\|_{H^{2\alpha}}} \left(\|\phi\|_{L^{d/\epsilon}} \|\mathbb{I}_{D_{k,n}}\|_{L^d/(d-\epsilon)}\right) \le C \|\mathbb{I}_{D_{k,n}}\|_{L^d/(d-\epsilon)}$$

where C is the constant resulting from the continuous embedding of $H^{d/2}(D)$ into $H^{d/2-\epsilon}(D)$. Thus, we obtain for all $\epsilon > 0$ the estimate

$$\|\mathbb{I}_{D_{k,n}}\|_{H^{-d/2}}^2 \le C^2 |D_{k,n}| v_+(n)^{(d-2\epsilon)/d}.$$

(b)Next, we have to establish that the Nemytzkii operator F on $L^2(D)$ is also Lipschitz continuous with respect to the norms $\|\cdot\|_{H^{-\alpha}}$, α≥0 , i.e., for all α≥0 there exists a constant $L_{-\alpha}$ such that

$$\left\|F(g_1,t) - F(g_2,t)\right\|_{H^{-\alpha}} \le L_{-\alpha}\|g_1 - g_2\|_{H^{-\alpha}} \quad \forall t \ge 0, g_1, g_2 \in L^2(D). \tag{4.16}$$

We obtain due to the Lipschitz continuity of f, which implies absolute continuity of f, that

$$\left|\int_D \phi(x)\big(F(g_1,t)(x) - F(g_2,t)(x)\big)\,\mathrm{d}x\right| = \left|\int_D \phi(x) \int_{z_1(t,x)}^{z_2(t,x)} f'(z)\,\mathrm{d}z\,\mathrm{d}x\right|.$$

where

$$z_1(t,x) = \int_D w(x,y) g_1(y)\,\mathrm{d}y + I(t,x),$$

$$z_2(t,x) = \int_D w(x,y) g_2(y)\,\mathrm{d}y + I(t,x).$$

Applying Hölder's inequality and the essential boundedness of the derivative f', we obtain the estimate

$$\left|\int_D \phi(x)\big(F(g_1,t)(x) - F(g_2,t)(x)\big)\,\mathrm{d}x\right|$$
$$\leq \|\phi\|_{L^p}\left(\int_D\left|\int_{z_1(t,x)}^{z_2(t,x)} f'(z)\,\mathrm{d}z\right|^q \mathrm{d}x\right)^{1/q}$$
$$\leq \|\phi\|_{L^p}\left(\int_D \big|\|f'\|_{L^\infty}\big(z_1(t,x) - z_2(t,x)\big)\big|^q \mathrm{d}x\right)^{1/q}$$
$$= \|\phi\|_{L^p}\|f'\|_{L^\infty}\left(\int_D\left|\int_D w(x,y)\big(g_1(y) - g_1(y)\big)\,\mathrm{d}y\right|^q \mathrm{d}x\right)^{1/q}.$$

Next, as by assumption $w(x,\cdot) \in H^\alpha(D).$, we obtain

$$\left(\int_D\left|\int_D w(x,y)\big(g_1(y) - g_1(y)\big)\,\mathrm{d}y\right|^q \mathrm{d}x\right)^{1/q}$$
$$= \left(\int_D \|w(x,\cdot)\|_{H^\alpha}^q \left|\left\langle w(x,\cdot)/\|w(x,\cdot)\|_{H^\alpha}, g_1 - g_2\right\rangle_{H^\alpha}\right|^q\right)^{1/q}$$
$$\leq \|w\|_{L^q\times H^\alpha}\|g_1 - g_2\|_{H^{-\alpha}}.$$

Overall, this yields the estimate

$$\left|\langle\phi, F(g_1,t) - F(g_2,t)\rangle_{H^{-\alpha}}\right| \leq \|\phi\|_{L^p}\left\|f'\right\|_{L^\infty}\|w\|_{L^q\times H^\alpha}\|g_1 - g_2\|_{H^{-\alpha}}.$$

Hence, taking the supremum on both sides of this inequality over all $\|\phi\|_{H^\alpha} = 1$, we obtain the Lipschitz condition (4.16) with $L_{-\alpha} := LK_\alpha\|w\|_{L^q\times H^\alpha}$, where K_α is the constant resulting from the continuous embedding of $H^\alpha(D)$ into $L^p(D)$ and the Lipschitz constant L of f satisfies $L \geq \|f'\|_{L^\infty}$.

4.3 Proof of Theorem 2.2 (Infinite Time Convergence)

(a)We first present an alternative representation for the jump processes $(\Theta_t^n)_{t\geq 0}$ and the solution v of the Wilson–Cowan equation (1.3). Using the generator of the PDMP $(\Theta_t^n, t)_{t\geq 0}$, we obtain that the components $\Theta^{k,n}$ satisfy

$$\Theta_t^{k,n} = \Theta_0^{k,n} + \int_0^t \lambda^n(\Theta_s^n, s)\int_{\mathbb{N}^p}\big(\xi^k - \Theta_s^{k,n}\big)\mu^n(\Theta_s^n, s; \mathrm{d}\xi)\,\mathrm{d}s + M_t^{k,n}$$
$$= \Theta_0^{k,n} + \int_0^t\left(-\frac{1}{\tau}\Theta_s^{k,n} + \frac{1}{\tau}l(k,n)\overline{f}_{k,n}(\Theta_s^n, s)\right)\mathrm{d}s + M_t^{k,n}. \quad (4 \tag{4.17}$$

where $(M_t^{k,n})_{t\geq 0}$ is a square-integrable càdlàg martingale given by

$$M_t^{k,n} := \Theta_t^{k,n} - \Theta_0^{k,n} - \int_0^t \lambda^n(\Theta_s^n, s) \int_{\mathbb{N}^p} (\xi^k - \Theta_s^{k,n}) \mu^n(\Theta_s^n, s; \mathrm{d}\xi)\,\mathrm{d}s. \tag{4.18}$$

As the jump process is regular, this martingale is almost surely of finite variation and it could also be written in terms of a stochastic integral with respect to the associated martingale measure of the PDMP [16]. Next, interpreting $\Theta^{k,n}$ as the solution of the stochastic evolution equation (4.17) driven by the martingale $M^{k,n}$, it follows from the variation of constants formula that it satisfies

$$\Theta_t^{k,n} = \mathrm{e}^{-t/\tau}\Theta_0^{k,n} + \frac{1}{\tau} l(k,n) \int_0^t \mathrm{e}^{-(t-s)/\tau} \overline{f}_{k,n}(\Theta_s^n, s)\,\mathrm{d}s$$
$$+ \int_0^t \mathrm{e}^{-(t-s)/\tau}\,\mathrm{d}M_s^{k,n}. \tag{4.19}$$

This formula can also be easily verified path-by-path by inserting (4.19) into (4.17) and using integration by parts. Note that here the stochastic integral with respect to the martingale is just a Riemann–Stieltjes integral as the martingale is of finite variation. For the sake of completeness, we briefly sketch the arguments. Thus, inserting (4.19) into (4.17) yields

$$\Theta_t^{k,n} = \underbrace{\Theta_0^{k,n} - \frac{1}{\tau}\int_0^t \mathrm{e}^{-s/\tau}\Theta_0^{k,n}\,\mathrm{d}s}_{(*)}$$
$$- \frac{1}{\tau} l(k,n)$$
$$\times \underbrace{\left(\frac{1}{\tau}\int_0^t \int_0^s \mathrm{e}^{-(s-r)/\tau}\overline{f}_{k,n}(\Theta_r^n, r)\,\mathrm{d}r\,\mathrm{d}s - \int_0^t \overline{f}_{k,n}(\Theta_s^n, s)\,\mathrm{d}s\right)}_{(**)}$$
$$\underbrace{-\frac{1}{\tau}\int_0^t \int_0^s \mathrm{e}^{-(s-r)/\tau}\,\mathrm{d}M_r^{k,n}\,\mathrm{d}s + M_t^{k,n}}_{(***)}.$$

Considering the three terms marked $(*)$–$(***)$separately, we show that this right-hand side equals (4.19). For the first term $(*)$ simply evaluating the integral yields

$$\Theta_0^{k,n} - \frac{1}{\tau}\int_{0t} e^{-s/\tau}\Theta_0^{k,n}\,\mathrm{d}s = \Theta_0^{k_n} - \frac{1}{\tau}(\alpha^{-1}e^{-t/\tau} - \tau)\Theta_0^{k,n} = e^{-t/\tau}\Theta_0^{k,n}.$$

which gives the first term in the right-hand side of (4.19). Next, we simplify the term $(**)$ employing integration by parts to the first term in $(**)$, which yields

$$\begin{aligned}
&\frac{1}{\tau}\int_0^t\int_0^s e^{-(s-r)/\tau}\overline{f}_{k,n}(\Theta_r^n, r)\,\mathrm{d}r\,\mathrm{d}s \\
&= \frac{1}{\tau}\int_0^t e^{(t-s)/\tau}\int_0^s e^{-(t-r)/\tau}\overline{f}_{k,n}(\Theta_r^n, r)\,\mathrm{d}r\,\mathrm{d}s \\
&= \frac{1}{\tau}\left(-\tau e^{(t-s)/\tau}\int_0^s e^{-(t-r)/\tau}\overline{f}_{k,n}(\Theta_r^n, r)\,\mathrm{d}r\right)\Bigg|_0^t \\
&\quad - \frac{1}{\tau}\int_0^t(-\tau)e^{(t-s)\tau}e^{-(t-s)/\tau}\overline{f}_{k,n}(\Theta_s^n, s)\,\mathrm{d}s \\
&= -\int_0^t e^{-(t-s)/\tau}\overline{f}_{k,n}(\Theta_s^n, s)\,\mathrm{d}s + \int_0^t \overline{f}_{k,n}(\Theta_s^n, s)\,\mathrm{d}s.
\end{aligned}$$

Thus, we obtain subtracting from this right-hand side the second term in

$$(**) = -\int_0^t e^{-(t-s)/\tau}\overline{f}_{k,n}(\Theta_s^n, s)\,\mathrm{d}s.$$

This term is just the second term in the right-hand side of (4.19). It remains to consider the term marked $(***)$. We have already stated that the stochastic integral with respect to the martingale (4.18) is defined path-by-path as a Riemann–Stieltjes integral, and thus satisfies

$$\begin{aligned}
&-\frac{1}{\tau}\int_0^s e^{-(s-r)/\tau}\,\mathrm{d}M_s^{k,n} \\
&= -\frac{1}{\tau}\sum_{\tau_j^n\le s} e^{-(s-\tau_j^n)/\tau}\left(\Theta_{\tau_j^n}^{k,n} - \Theta_{\tau_{j-}^n}^{k,n}\right) \\
&\quad + \frac{1}{\tau}\int_0^s e^{-(s-r)/\tau}\lambda^n(\Theta_r^n, r) \\
&\quad \times \int_{\mathbb{N}_0^p}(\xi^k - \Theta_r^{k,n})\mu^n((\Theta_r^n, r), \mathrm{d}\xi)\,\mathrm{d}s,
\end{aligned} \tag{4.20}$$

where τ_j^n denotes the j th jump time of the n th PDMP. Integrating the sum in this right-hand side over $(0, t)$ yields

$$-\frac{1}{\tau}\int_0^t \sum_{\tau_j^n \le s} \mathrm{e}^{-(s-\tau_j^n)/\tau}\left(\Theta_{\tau_j^n}^{k,n} - \Theta_{\tau_{j-}^n}^{k,n}\right)\mathrm{d}s$$

$$= \sum_{\tau_j^n \le t} \mathrm{e}^{-(t-\tau_j^n)/\tau}\left(\Theta_{\tau_j^n}^{k,n} - \Theta_{\tau_{j-}^n}^{k,n}\right) - \sum_{\tau_j^n \le t}\left(\Theta_{\tau_j^n}^{k,n} - \Theta_{\tau_{j-}^n}^{k,n}\right)$$

$$= \sum_{\tau_j^n \le t} \mathrm{e}^{-(t-\tau_j^n)/\tau}\left(\Theta_{\tau_j^n}^{k,n} - \Theta_{\tau_{j-}^n}^{k,n}\right) - \left(\Theta_t^{k,n} - \Theta_0^{k,n}\right).$$

Next, we apply integration by parts to the integral over $(0, t)$ of the second term above analogously to the application to term ($(**)$, and obtain

$$\frac{1}{\tau}\int_0^t \int_0^s \mathrm{e}^{-\alpha(s-r)}\lambda^n(\Theta_r^n, r)\int_{\mathbb{N}_0^P}(\xi^k - \Theta_r^{k,n})\mu^n\big((\Theta_r^n, r), \mathrm{d}\xi\big)\,\mathrm{d}r\,\mathrm{d}s$$

$$= -\int_0^t \mathrm{e}^{-(t-s)/\tau}\lambda^n(\Theta_s^n, s)\int_{\mathbb{N}_0^P}(\xi^k - \Theta_s^{k,n})\mu^n\big((\Theta_s^n, s), \mathrm{d}\xi\big)\,\mathrm{d}s$$

$$+ \int_0^t \lambda^n(\Theta_s^n, s)\int_{\mathbb{N}_0^P}(\xi^k - \Theta_s^{k,n})\mu^n\big((\Theta_s^n, s), \mathrm{d}\xi\big)\,\mathrm{d}s.$$

Hence, overall these considerations show that

$$(***) = \int_0^t \mathrm{e}^{-(t-s)/\tau}\,\mathrm{d}M_s^{k,n}$$

and we obtain the final, third term in the right-hand side of (4.19). This completes the proof that (4.19) solves Eq. (4.17).

Further, we obtain from the variation of constants formula for $\Theta_t^{k,n}$ also a representation for the stochastic mean activity v n vnν^n by inserting (4.19) into its definition (2.7). This gives

$$\nu_t^n = \mathrm{e}^{-t/\tau}\nu_0^n + \frac{1}{\tau}\sum_{k=1}^{P}\int_0^t \mathrm{e}^{-(t-s)/\tau}\overline{f}_{k,n}(\Theta_s^n, s)\,\mathrm{d}s\,\mathbb{I}_{D_{k,n}}$$

$$+\sum_{k=1}^{P}\frac{1}{l(k,n)}\int_0^t \mathrm{e}^{-(t-s)/\tau}\,\mathrm{d}M_s^{k,n}\mathbb{I}_{D_{k,n}}$$

$$= e^{-t/\tau} v_0^n + \frac{1}{\tau} \int_0^t e^{-(t-s)/\tau} \overline{F}^n (v_s^n, s)\, ds$$

$$+ \sum_{k=1}^{P} \frac{1}{l(k,n)} \int_0^t e^{-(t-s)/\tau}\, dM_s^{k,n} \mathbb{I}_{D_{k,n}}. \tag{4.21}$$

Finally, in order to compare stochastic and deterministic solutions we use that the solution of the Wilson–Cowan equation can also be given via the variation of constants formula, i.e., it holds that for all $t \geq 0$

$$v(t) = e^{-t/\tau} v(0) + \frac{1}{\tau} \int_0^t e^{-(t-s)/\tau} F\big(v(s), s\big)\, ds. \tag{4.22}$$

Thus, subtracting (4.22) from (4.21), and taking the expectation of the norm in $H^{-\alpha}(D)$ yields the estimate

$$\mathbb{E}^n \left\| v(t) - v_t^n \right\|_{H^{-\alpha}}$$

$$= e^{-t/\tau} \mathbb{E}^n \left\| v(0) - v_0^n \right\|_{H^{-\alpha}}$$

$$+ \frac{1}{\tau} \int_0^t e^{-(t-s)/\tau} \mathbb{E}^n \left\| F\big(v(s), s\big) - \overline{F}^n (v_s^n, s) \right\|_{H^{-\alpha}}$$

$$+ \mathbb{E}^n \left\| \sum_{k=1}^{P} \frac{1}{l(k,n)} \int_0^t e^{-(t-s)/\tau}\, dM_s^{k,n} \mathbb{I}_{D_{k,n}} \right\|_{H^{-\alpha}}. \tag{4.23}$$

We deal with the terms in the right-hand side of (4.23) separately in the following such that we can apply Gronwall's inequality. Note that the term containing the initial condition vanishes due to the assumptions of the theorem. We start with the stochastic integrals in the subsequent part (b) of the proof.

(b)As due to Jensen's inequality $\mathbb{E}|Y| \leq \sqrt{\mathbb{E}|Y|^2}$, it makes sense to calculate the second moment of the stochastic integral in the right-hand side. For the norm in $H^{-\alpha}(D)$, we use $\|\phi\|_{H^{-\alpha}}^2 = (\phi, \phi)_{H^{-\alpha}}$, and thus obtain using the linearity of the inner product

$$\left\| \sum_{k=1}^{P} \frac{1}{l(k,n)} \underbrace{\int_0^t e^{-(t-s)/\tau} \, dM_s^{k,n} \, \mathbb{I}_{D_{k,n}}}_{=:\beta_t^{k,n}} \right\|_{H^{-\alpha}}^2$$

$$= \sum_{k=1}^{P} \frac{|\beta_{k,n}|^2}{l(k,n)^2} \|\mathbb{I}_{D_{k,n}}\|_{H^{-\alpha}}^2 + \sum_{\substack{k,j=1 \\ k \neq j}}^{P} \frac{\beta_{k,n}\beta_{j,n}}{l(k,n)l(j,n)} (\mathbb{I}_{D_{k,n}}, \mathbb{I}_{D_{j,n}})_{H^{-\alpha}}.$$

We next consider the individual expectations of the random terms $|\beta_{k,n}|^2$ and $\beta_{k,n}\beta_{j,n}$ in the right-hand side. We have already stated that the stochastic integral with respect to the martingale (4.18) is defined path-by-path as a Riemann–Stieltjes integral (see (4.20)) and, moreover, (4.20) implies that the stochastic convolution integral can be written as a stochastic integral with respect to the fundamental martingale measure M^n associated with the PDMP $(\Theta_t^n, t)_{t\geq 0}$ (see [16]), i.e.,

$$\int_0^t e^{-(t-s)/\tau} \, dM_s^{k,n} = \int_{[0,t]\times\mathbb{N}_0^P} e^{-(t-s)/\tau} \left(\xi^k - \Theta_{s-}^{k,n}\right) M^n(ds, d\xi)$$

with predictable integrand

$$(\xi, s, \omega) \mapsto e^{-(t-s)/\tau} \left(\xi^k - \Theta_{s-}^{k,n}(\omega)\right).$$

Then we obtain due to the Itô-isometry following from Proposition 4.6.2 in [16] using (4.6) that

$$\mathbb{E}^n \left| \int_0^t e^{-(t-s)/\tau} \, dM_s^{k,n} \right|^2$$

$$= \mathbb{E}^n \int_0^t \lambda^n(Y_s^n) \int_{\mathbb{N}_0^P} e^{-2(t-s)/\tau} \left(\xi^k - \Theta_s^{k,n}\right)^2 \mu^n(Y_s^n, d\xi) \, ds$$

$$\leq \mathbb{E}^n \int_0^t e^{-2(t-s)/\tau} \left(\frac{1}{\tau}\Theta_s^{k,n} + \frac{1}{\tau} l(k,n) \overline{f}_{k,n}(Y_s^n) \right) ds$$

$$\leq \frac{1}{\tau}\left(\mathbb{E}^n \Theta_0^{k,n} + 2l(k,n)\|f\|_0\right) \int_0^t e^{-2(t-s)/\tau} \, ds.$$

It remains to consider the product $\beta_t^{k,n}\beta_t^{j,n}$ for which we obtain due to the integration by parts formula

$$\beta_t^{k,n}\beta_t^{j,n} = \int_0^t \beta_{s-}^{k,n}\,\mathrm{d}\beta_s^{j,n} + \int_0^t \beta_{s-}^{j,n}\,\mathrm{d}\beta_s^{k,n} + \left[\beta^{k,n}, \beta^{j,n}\right]_t, \tag{4.24}$$

where the square brackets denote the quadratic variation process. The expectation of each of the terms in the right-hand side vanishes: The first two are stochastic integrals with respect to martingales, hence martingales themselves which are identical to zero at the origin. Furthermore, as both martingales are càdlàg with paths of finite variation on compacts, hence quadratic pure jump martingales, we obtain for the quadratic variation process

$$\left[\beta^{k,n}, \beta^{j,n}\right]_t = \sum_{s\le t}\left(\beta_s^{k,n} - \beta_{s-}^{k,n}\right)\left(\beta_s^{j,n} - \beta_{s-}^{j,n}\right).$$

However, as all jump times of the two martingales a.s. differ it follows that $[\beta^{k,n}, \beta^{j,n}]_t = 0$.

Thus, overall we have established that

$$\begin{aligned}\mathbb{E}^n\left\|\sum_{k=1}^P \frac{1}{l(k,n)}\int_0^t \mathrm{e}^{-(t-s)/\tau}\,\mathrm{d}M_s^{k,n}\mathbb{I}_{D_{k,n}}\right\|_{H^{-\alpha}}^2 \\ = \frac{1}{2}\sum_{k=1}^P \frac{1+2\|f\|_0}{l(k,n)}\|\mathbb{I}_{D_{k,n}}\|_{H^{-\alpha}}^2,\end{aligned} \tag{4.25}$$

where $1/2$ is an upper bound for $\frac{1}{\tau}\int_0^t \mathrm{e}^{-2(t-s)/\tau}$ independent of t. Estimating the norm $\|\mathbb{I}_{D_{k,n}}\|_{H^{-\alpha}}^2$ just as in the proof of Corollary 2.1, we finally obtain that

$$\mathbb{E}^n\left\|\sum_{k=1}^P \frac{1}{l(k,n)}\int_0^t \mathrm{e}^{-(t-s)/\tau}\,\mathrm{d}M_s^{k,n}\mathbb{I}_{D_{k,n}}\right\|_{H^{-\alpha}} \le \frac{1}{4}\left(\left(1+2\|f\|_0\right)|D|\frac{v_+(n)^r}{\ell_-(n)}\right)^{1/2}.$$

with
$r = 2\alpha/d$ for $0 \le \alpha \le d/2$, $r = 1-\epsilon$ for $\alpha = d/2$ and $r = 1$ for $\alpha > d/2$.

(c)We next estimate the term

$$\int_0^t \mathrm{e}^{-(t-s)/\tau}\mathbb{E}^n\left\|F\left(v(s), s\right) - \overline{F}^n\left(v_s^n, s\right)\right\|_{H^{-\alpha}}$$

in (4.23). From part (b) of the proof of Theorem 2.1 in Sect. 4.1, it follows that

$$\mathbb{E}^n \left\| F(v_t^n, t) - \overline{F}^n(v_t^n, t) \right\|_{H^{-\alpha}}$$
$$\leq \delta_+(n) \frac{K_{-\alpha} L}{\pi} \left(\sqrt{|D|}(1 + \|f\|_0) \|\nabla_x w\|_{L^2 \times L^2} + \left\| \nabla_x I(t) \right\|_{L^2} \right). \tag{4.26}$$

where F is the Nemytzkii operator defined in (4.4) and $K_{-\alpha}$ is a constant resulting from the continuous embedding of $L^2(D)$ into $H^{-\alpha}(D)$. Here, the right-hand side can be further estimated independently of $t \geq 0$ using the assumption that $\|\nabla_x I(t)\|_{L^2}$ is uniformly bounded in $t \geq 0$. Furthermore, we have shown in Sect. 4.2 in the proof of Corollary 2.1, that under the appropriate assumptions the Nemytzkii operator F is Lipschitz continuous on $H^{-\alpha}(D)$, $\alpha \geq 0$, with Lipschitz constant $L_{-\alpha} > 0$ independent of $t \geq 0$, i.e.,

$$\left\| F(g_1, t) - F(g_2, t) \right\|_{H^{-\alpha}} \leq L_{-\alpha} \|g_1 - g_2\|_{H^{-\alpha}} \quad \forall g_1, g_2 \in L^2(D). \tag{4.27}$$

A combination of the triangle inequality and the estimates (4.26) and (4.27) yields

$$\int_0^t e^{-(t-s)/\tau} \mathbb{E}^n \left\| F(v(s), s) - \overline{F}^n(v_s^n, s) \right\|_{H^{-\alpha}}$$
$$\leq L_{-\alpha} \int_0^t e^{-(t-s/\tau)} \mathbb{E}^n \left\| v(s) - v_s^n \right\|_{H^{-\alpha}} ds + \mathcal{O}\big(\delta_+(n)\big).$$

Overall, it thus follows from (4.23) that

$$\mathbb{E}^n \left\| v(t) - v_t^n \right\|_{H^{-\alpha}} \leq \mathbb{E}^n \left\| v(0) - v_0^n \right\|_{H^{-\alpha}}$$
$$+ \frac{L_{-\alpha}}{\tau} \int_0^t e^{-(t-s/\tau)} \mathbb{E}^n \left\| v(s) - v_s^n \right\|_{H^{-\alpha}} ds$$
$$+ \mathcal{O}\left(\delta_+(n) + \sqrt{\frac{v_+(n)^r}{\ell_-(n)}} \right).$$

Then an application of Gronwall's inequality yields

$$\mathbb{E}^n \left\| v(t) - v_t^n \right\|_{H^{-\alpha}} \leq \left(\mathbb{E}^n \left\| v(0) - v_0^n \right\|_{H^{-\alpha}} + \mathcal{O}\left(\delta_+(n) + \sqrt{\frac{v_+(n)^r}{\ell_-(n)}} \right) \right)$$
$$\times \exp\left(\frac{L_{-\alpha}}{\tau} \int_0^t e^{-(t-s)/\tau} ds \right)$$
$$\leq \left(\mathbb{E}^n \left\| v(0) - v_0^n \right\|_{H^{-\alpha}} + \mathcal{O}\left(\delta_+(n) + \sqrt{\frac{v_+(n)^r}{\ell_-(n)}} \right) \right) e^{L_{-\alpha}}.$$

By assumptions of the theorem, the term in the right-hand side converges to zero for n→∞ n→∞$n \to \infty$. As this convergence is uniform in t, it holds that

$$\lim_{n\to\infty} \sup_{t\geq 0} \mathbb{E}^n \left\| \nu(t) - \nu_t^n \right\|_{H^{-\alpha}} = 0.$$

(4.28)

4.4 Proof of Theorem 2.3 (Martingale Central Limit Theorem)

In order to prove the martingale central limit theorem, we employ the general martingale central limit theorem (Theorem 5.1 in [27]) for the Hilbert space H −α (D) H−α(D)$H^{-\alpha}(D)$, i.e., the dual of the Sobolev space Hα(D), for α>d. To apply this theorem, it suffices to prove the following conditions. Subsequently we use $\rho_n = \sqrt{\ell_-(n)/\nu_+(n)}$ to denote the re-scaling sequence and use the notation

$$\left\langle G^n(t)\phi, \phi \right\rangle_{H^\alpha} = \lambda\left(Y_t^n\right) \int_{\mathbb{N}_0^P} \left\langle \nu^n(\xi) - \nu^n\left(\Theta_t^n\right), \phi \right\rangle_{H^\alpha}^2 \mu^n\left(Y_t^n, \mathrm{d}\xi\right) \tag{4.29}$$

which corresponds to the quadratic variation of the martingales $(M_t^n)_{t\geq 0}$, see [27] for a discussion.

(CLT1) For $t > 0$, it holds that

$$\sup_{n\in\mathbb{N}} \rho_n \mathbb{E}^n \int_0^t \left[\lambda^n\left(Y_s^n\right) \int_{\mathbb{N}^P} \left\| \nu^n(\xi) - \nu^n\left(\Theta_s^n\right) \right\|_{H^{-\alpha}}^2 \mu^n\left(Y_s^n, \mathrm{d}\xi\right) \mathrm{d}s \right] < \infty, \tag{4.30}$$

and there exists an orthonormal basis $(\varphi_j)_{j\in\mathbb{N}}$ of $H^\alpha(D)$ such that for all $j \in \mathbb{N}$ and $t \geq 0$

$$\rho_n \mathbb{E}^n \left\langle G^n\left(Y_t^n\right)\varphi_j, \varphi_j \right\rangle_{H^\alpha} \leq \gamma_j C,$$

(4.31)

where the constants $\gamma_j > 0$ are independent of n and t, satisfy $\sum_{j\geq 1} \gamma_j < \infty$, and the constant C>0 C>0$C > 0$ is independent of n and k but may depend on t.

(CLT2) The jump heights of the re-scaled martingales are almost surely uniformly bounded, i.e., there exists a constant $\beta < \infty$ such that it holds almost surely for all $n \in \mathbb{N}$ that

$$\sup_{t\geq 0} \sqrt{\rho_n} \left\| \nu^n\left(\Theta_t^n\right) - \nu^n\left(\Theta_{t-}^n\right) \right\|_{H^{-\alpha}} < \beta. \tag{4.32}$$

Further, for all ϕ∈H α (D) ϕ∈Hα(D)$\phi \in H^{\alpha}(D)$ and all t>0 t>0$t > 0$ it holds that

$$\lim_{n\to\infty} \int_0^t \mathbb{E}^n \left| \langle G(v(s))\phi, \phi\rangle_{H^\alpha} - \rho_n \langle G^n(Y_s^n)\phi, \phi\rangle_{H^\alpha} \right| \mathrm{d}s = 0. \tag{4.33}$$

On a technical level, we note that the condition (CLT1) guarantees tightness of the sequence of re-scaled martingales $(\sqrt{\rho_n} M_t^n)_{t\geq 0}$ in the Skorokhod space of càdlàg functions in $H^{-\alpha}(D)$. This property is equivalent to relative compactness in the topology of weak convergence of measures, and thus implies the existence of a convergent subsequence. The conditions (CLT2) are then sufficient to establish that any limit possesses the form of a diffusion process defined by the covariance operator C given in (2.18). In particular, condition (4.33) precisely gives the convergence of the quadratic variations and is thus the central condition. In the subsequent two parts of the proof, we show that they are satisfied: In part (a), we prove conditions (4.30) and (4.31), and part (b) establishes (4.32) and (4.33).

(a)We first prove conditions (4.30) and (4.31). Here, we also observe the significance of the choice of the norm in H −α (D) H−α(D)$H^{-\alpha}(D)$ with α>d α>d$\alpha > d$ for establishing the convergence, which is essentially that it guarantees the existence a Sobolev space H α 1 (D) Hα1(D)$H^{\alpha_1}(D)$ with continuous embeddings $H^{\alpha}(D) \hookrightarrow H^{\alpha_1}(D) \hookrightarrow C(\overline{D})$, where the first is of Hilbert–Schmidt type. For subsequent use, we recall the estimates

$$\|\mathbb{I}_{D_{k,n}}\|_{H^{-\alpha}}^2 \leq K_\alpha^2 |D_{k,n}|^2$$

with a suitable constant $K_\alpha > 0$, which we have already established in the proof of Corollary 2.1 due to the Hölder inequality and the Sobolev embedding theorem. Therefore, we obtain for the term inside the expectation in (4.30) the estimate

$$\lambda^n(Y_s^n) \int_{\mathbb{N}_0^P} \| v^n(\xi) - v^n(\Theta_s^n) \|_{H^{-\alpha}}^2 \mu^n(Y_s^n, \mathrm{d}\xi)$$

$$\leq \frac{1}{\tau} K_\alpha^2 \sum_{k=1}^{P} \frac{|D_{k,n}|^2}{l(k,n)^2} \left(\Theta_s^{k,n} + l(k,n) \overline{f}_{k,n}(Y_s^n) \right).$$

Next taking the expectation, using the bound (4.6) on $\mathbb{E}^n \Theta_s^{k,n}$ and integrating over $[0,t]$ we obtain the estimate

$$\int_0^t \mathbb{E}^n\left[\lambda^n(Y_s^n)\int_{\mathbb{N}_0^P}\|v^n(\xi)-v^n(\Theta_s^n)\|_{H^{-\alpha}}^2\mu^n(Y_s^n,\mathrm{d}\xi)\right]\mathrm{d}s$$
$$\leq \frac{t}{\tau}K_\alpha^2(1+2\|f\|_0)\frac{v_+(n)}{\ell_-(n)}.$$

Multiplying both sides with $\rho_n = \ell_-(n)/v_+(n)$, we find that condition (4.30) is satisfied.

We proceed to condition (4.31) and first of all expand the integrand to obtain

$$\langle G^n(Y_s^n)\varphi_j,\varphi_j\rangle_{H^\alpha} = \lambda^n(Y_s^n)\int_{\mathbb{N}_0^P}\langle v^n(\xi)-v^n(\Theta_s^n),\phi_j\rangle_{H^{-\alpha}}^2\mu^n(Y_s^n,\mathrm{d}\xi)$$
$$= \frac{1}{\tau}\sum_{k=1}^P\frac{1}{l(k,n)^2}(\Theta_s^{k,n}+l(k,n)\overline{f}_{k,n}(Y_s^n))\langle \mathbb{I}_{D_{k,n}},\varphi_j\rangle_{H^\alpha}^2.$$

We next estimate the term $\langle \mathbb{I}_{D_{k,n}},\varphi\rangle_{H^\alpha}^2$. Here, we use the fact that for a function in $L^2(D)$ its application as an element of the dual H -α (D) H−α(D)$H^{-\alpha}(D)$ as well as $H^{-\alpha_1}(D)$ for any α 1 α_1 with $0\leq\alpha_1<\alpha$ coincide. We choose α 1 α1α_1 such that $d/2<\alpha_1<\alpha-d/2$ and obtain

$$\langle \mathbb{I}_{D_{k,n}},\varphi\rangle_{H^\alpha}^2 \leq \|\mathbb{I}_{D_{k,n}}\|_{H^{-\alpha_1}}^2\|\varphi_j\|_{H^{\alpha_1}}^2 \leq K_{\alpha_1}^2|D_{k,n}|^2\|\varphi_j\|_{H^{\alpha_1}}^2,$$

where K_{α_1} is the constant resulting from the Sobolev embedding theorem. Next, taking the expectation, estimating the expectation terms as before and multiplying by ρ n ρnρ_n yields

$$\rho_n\mathbb{E}^n\langle G^n(Y_t^n)\varphi_j,\varphi_j\rangle_{H^\alpha} \leq \frac{1}{\tau}K_{\alpha_1}^2(1+\|f\|_0)\|\varphi_j\|_{H^{\alpha_1}}^2.$$

We chose the constants in (4.31) as $C := K_{\alpha_1}^2(1+\|f\|_0)/\tau$ and $\gamma_j := \|\varphi_j\|_{H^{\alpha_1}}^2$. Finally, as due to Maurin's theorem the embedding of the space $H^\alpha(D)$ into $H^{\alpha_1}(D)$ is of Hilbert–Schmidt type, cf. footnote 3 on page 16, it holds that $\sum_{j\geq 1}\|\varphi_j\|_{H^{\alpha_1}}^2<\infty$. Condition (4.31) is satisfied.(b)

The estimates in part (a) further show that the jump sizes are almost surely uniformly bounded as

$$\sup_{t\geq 0}\sqrt{\rho_n}\left\|v^n(\Theta_t^n)-v^n(\Theta_{t-}^n)\right\|_{H^{-\alpha}} \leq K_\alpha\sqrt{\frac{v_+(n)}{\ell_-(n)}}.$$

Here, the upperbound in the right-hand side converges to zero for $n\to\infty$, and thus the left-hand side is bounded over all $n\in\mathbb{N}$. Therefore, condition (4.32) holds and we are left to prove the convergence of the quadratic variation (4.33). For the jump process, the quadratic variation satisfies

$$\begin{aligned}\langle G^n(t)\phi,\phi\rangle_{H^\alpha} &= \lambda(Y_t^n)\int_{\mathbb{N}_0^P}\langle v^n(\xi)-v^n(\Theta_t^n),\phi\rangle_{H^\alpha}^2\mu^n(Y_t^n,\mathrm{d}\xi)\\ &= \frac{1}{\tau}\sum_{k=1}^{P}\frac{1}{l(k,n)^2}\left(\Theta_t^{k,n}+l(k,n)\overline{f}_{k,n}(Y_t^n)\right)\langle\mathbb{I}_{D_{k,n}},\phi\rangle_{H^\alpha}^2.\end{aligned}$$

The quadratic variation of the limiting diffusion is given by

$$\begin{aligned}&\langle G(v(t),t)\phi,\phi\rangle_{H^\alpha}\\ &= \int_D\phi(x)^2\left(\frac{1}{\tau}v(t,x)+\frac{1}{\tau}f\left(\int_D w(x,y)v(t,y)\,\mathrm{d}y+I(t,x)\right)\right)\mathrm{d}x.\end{aligned}$$

Here, the necessary estimates are split into several parts which are separately considered in the following. Afterwards, the estimates are combined to infer the convergence (4.33). In the following, we use again F as the Nemytzkii operator defined in (4.4). Hence, for the difference of the quadratic variations, we obtain the estimate

$$\begin{aligned}&\mathbb{E}^n\left|\langle G(v(t),t)\phi,\phi\rangle_{H^\alpha}-\rho_n\langle G^n(t)\phi,\phi\rangle_{H^\alpha}\right|\\ &= \frac{1}{\tau}\mathbb{E}^n\left|\int_D\phi(x)^2v(t,x)+\phi(x)^2F(v(t),t)(x)\,\mathrm{d}x\right.\\ &\quad\left.-\sum_{k=1}^{P}\frac{\rho_n}{l(k,n)^2}\left(\Theta_t^{k,n}+l(k,n)\overline{f}_{k,n}(Y_t^n)\right)\langle\mathbb{I}_{D_{k,n}},\phi\rangle_{H^\alpha}^2\right|\\ &\le\frac{1}{\tau}\mathbb{E}^n\left|\int_D\underbrace{\phi(x)^2v(t,x)}_{\text{(i)}}+\underbrace{\phi(x)^2F(v(t),t)(x)}_{\text{(ii)}}\,\mathrm{d}x\right.\\ &\quad\left.-\int_D\underbrace{\phi(x)^2v^n(\Theta_t^n)(x)}_{\text{(i)}}+\underbrace{\phi(x)^2F(v^n(\Theta_t^n),t)(x)}_{\text{(ii)}}\,\mathrm{d}x\right|\end{aligned}\qquad(4.34)$$

Using the triangle inequality once again for each of the two differences grouping the terms marked (i)–(iv) we obtain four terms, which we subsequently estimate separately. Finally, in part (v), we combine the four estimates. (i)

The first term is the simplest to estimate. Using the Cauchy–Schwarz inequality, we obtain

$$\mathbb{E}^n\left|\int_D \phi^2(x)\big(v(t,x)-v_t^n(x)\big)\,\mathrm{d}x\right| \le \|\phi\|_{L^4}^2\mathbb{E}^n\left\|v(t)-v_t^n\right\|_{L^2}. \tag{4.35}$$

(ii)We next consider the difference arising from the terms marked (ii) and obtain using the Lipschitz condition (1.4) on f and the Cauchy–Schwarz inequality twice

$$\begin{aligned}
&\mathbb{E}^n\left|\int_D \phi(x)^2\big(F\big(v(t),t\big)(x)-F\big(v_t^n,t\big)(x)\big)\,\mathrm{d}x\right| \\
&\le L\mathbb{E}^n\int_D |\phi(x)|^2\left|\int_D w(x,y)\big(v(t,y)-v_t^n(y)\big)\,\mathrm{d}y\right|\mathrm{d}x \\
&\le L\mathbb{E}^n\int_D |\phi(x)|^2\left\|w(x,\cdot)\right\|_{L^2}\left\|v(t)-v_t^n\right\|_{L^2}\mathrm{d}x \\
&\le L\|\phi\|_{L^4}^2\|w\|_{L^2\times L^2}\mathbb{E}^n\left\|v(t)-v_t^n\right\|_{L^2}.
\end{aligned} \tag{4.36}$$

(iii)In order to estimate the next term, we use the bound (4.6) on $\mathbb{E}^n\Theta_t^{k,n}$, and thus obtain

$$\begin{aligned}
&\mathbb{E}^n\sum_{k=1}^P \frac{\Theta_t^{k,n}}{l(k,n)}\left|\int_{D_{k,n}}\phi(x)^2\,\mathrm{d}x-\frac{\rho_n}{l(k,n)}\left(\int_{D_{k,n}}\phi(x)\,\mathrm{d}x\right)^2\right| \\
&\le (1+\|f\|_0) \\
&\quad\times\sum_{k=1}^P |D_{k,n}|\left|\frac{1}{|D_{k,n}|}\int_{D_{k,n}}\phi(x)^2\,\mathrm{d}x-\left(\frac{1}{|D_{k,n}|}\int_{D_{k,n}}\phi(x)\,\mathrm{d}x\right)^2\right| \\
&\quad+(1+\|f\|_0)\sum_{k=1}^P |D_{k,n}|\left|1-\frac{\rho_n|D_{k,n}|^2}{l(k,n)|D_{k,n}|}\right|\left(\frac{1}{|D_{k,n}|}\int_{D_{k,n}}\phi(x)\,\mathrm{d}x\right)^2 \\
&\le (1+\|f\|_0)\sum_{k=1}^P\int_{D_{k,n}}\left(\phi(x)-\frac{1}{|D_{k,n}|}\int_{D_{k,n}}\phi(y)\,\mathrm{d}y\right)^2\mathrm{d}x \\
&\quad+(1+\|f\|_0)\left|1-\frac{v_-(n)}{v_+(n)}\frac{\ell_-(n)}{\ell_+(n)}\right|\sum_{k=1}^P|D_{k,n}|\left(\frac{1}{|D_{k,n}|}\int_{D_{k,n}}\phi(x)\,\mathrm{d}x\right)^2.
\end{aligned}$$

Then the estimate is completed applying the Poincaré inequality (4.1) to the first term, that is, estimating

$$\sum_{k=1}^P\int_{D_{k,n}}\left(\phi(x)-\frac{1}{|D_{k,n}|}\int_{D_{k,n}}\phi(y)\,\mathrm{d}y\right)^2\mathrm{d}x \le \frac{\mathrm{diam}(D_{k,n})^2}{\pi^2}\|\nabla\phi\|_{L^2}^2.$$

and the observation that the second term is proportional to $\|\overline{\phi}^n\|_{L^2}^2$, which is the piecewise constant approximation to ϕ based on the partition $\mathcal{D}_n$; see (4.2). Therefore, we overall obtain an upper bound for the difference constituted by the terms (iii) in (4.34) by

$$\mathbb{E}^n\left|\int_D \phi(x)^2 v_t^n(x)\,\mathrm{d}x - \rho_n \sum_{k=1}^{P} \frac{\Theta_t^{k,n}}{l(k,n)^2}\langle \mathbb{I}_{D_{k,n}}, \phi\rangle_{H^\alpha}^2\right|$$
$$\leq \delta_+(n)^2 \frac{1+\|f\|_0}{\pi^2}\|\phi\|_{H^1}^2 + \big(1+\|f\|_0\big)R(n). \tag{4.37}$$

In the last term,

$$R(n) := \left|1 - \frac{v_-(n)}{v_+(n)}\frac{\ell_-(n)}{\ell_+(n)}\right| \|\overline{\phi}^n\|_{L^2}^2$$

converges to zero for $n \to \infty$ by assumption (2.19) and as the sequence $\|\overline{\phi}^n\|_{L^2}$ is bounded as it converges to $\|\phi\|_{L^2}$ for $n \to \infty$.(iv)

Finally, we consider the difference

$$\mathbb{E}^n\left|\int_D \phi(x)^2 F(v_t^n, t)(x)\,\mathrm{d}x - \rho_n \sum_{k=1}^{P} \frac{l(k,n)}{l(k,n)^2}\overline{f}_{k,n}(Y_t^n)\langle \mathbb{I}_{D_{k,n}}, \phi\rangle_{H^\alpha}^2\right|$$
$$\leq \mathbb{E}^n \sum_{k=1}^{P}\left|\int_{D_{k,n}} \phi(x)^2 F(v_t^n, t)(x)\,\mathrm{d}x - \frac{\rho_n}{l(k,n)}\overline{f}_{k,n}(Y_t^n)\langle \phi, \mathbb{I}_{D_{k,n}}\rangle_{H^\alpha}^2\right|.$$

We continue estimating the difference in each summand in the final right-hand side and obtain using the triangle inequality for the term inside the expectation

$$\mathbb{E}^n \sum_{k=1}^{P}\left|\int_{D_{k,n}} \phi(x)^2 F(v_t^n, t)(x)\,\mathrm{d}x - \frac{\rho_n}{l(k,n)}\overline{f}_{k,n}(Y_t^n)\langle \mathbb{I}_{D_{k,n}}, \phi\rangle_{H^\alpha}^2\right|$$
$$\leq \underbrace{\mathbb{E}^n \sum_{k=1}^{P}\left|\int_{D_{k,n}} \phi(x)^2\big(F(v_t^n, t)(x) - \overline{f}_{k,n}(Y_t^n)\big)\,\mathrm{d}x\right|}_{(*)}$$
$$+ \underbrace{\mathbb{E}^n \sum_{k=1}^{P}\left|\overline{f}_{k,n}(Y_t^n)\left(\int_{D_{k,n}} \phi(x)^2\,\mathrm{d}x - \frac{\rho_n}{l(k,n)}\langle \mathbb{I}_{D_{k,n}}, \phi\rangle_{H^\alpha}^2\right)\right|}_{(**)}.$$

We start with the first term and observe that it possesses the same structure as the term estimated in part (c) of the proof of Theorem 2.1 with the only difference that here the function ϕ in the integrand is squared. Therefore, we obtain the estimate, cf. (4.15),

$$(*) \leq \delta_+(n) \frac{L\|\phi\|_{L^4}^2}{\tau\pi} \big(\sqrt{|D|}(1 + \|f\|_0)\|\nabla_x w\|_{L^2 \times L^2} + \big\|\nabla_x I(t)\big\|_{L^2}\big).$$

Next, we estimate the second term. Note that f ¯ k,n f¯k,n $\overline{f}_{k,n}$ is bounded by ||f|| 0 ||f||0 $\|f\|_0$, and thus the remaining term is just as in part (iii) of the proof. Hence, we obtain the estimate, cf. (4.37),

$$(**) \leq \delta_+(n)^2 \frac{\|f\|_0}{\pi^2}\|\phi\|_{H^1}^2 + \|f\|_0 \left|1 - \frac{v_-(n)}{v_+(n)} \frac{\ell_-(n)}{\ell_+(n)}\right| \|\overline{\phi}^n\|_{L^2}^2.$$

Therefore, we overall obtain an upper bound for the difference generated by the terms (iv) by

$$\begin{aligned} &\mathbb{E}^n \left| \int_D \phi(x)^2 F(v_t^n, t)(x)\,\mathrm{d}x - \rho_n \sum_{k=1}^{P} \frac{1}{l(k,n)} \overline{f}_{k,n}(Y_t^n) \langle \mathbb{I}_{D_{k,n}}, \phi\rangle_{H^\alpha}^2 \right| \\ &\leq \delta_+(n) \frac{L\|\phi\|_{L^4}^2}{\pi} \big(\sqrt{|D|}(1 + \|f\|_0)\|\nabla_x w\|_{L^2 \times L^2} + \big\|\nabla_x I(t)\big\|_{L^2}\big) \\ &\quad + \delta_+(n)^2 \frac{\|f\|_0}{\pi^2}\|\phi\|_{H^1}^2 + \|f\|_0 R(n), \end{aligned} \tag{4.38}$$

where the term R(n) R(n)$R(n)$ is as in (4.37).

(v)To complete the proof, we combine the estimates (4.35)–(4.38) to obtain

$$\begin{aligned} &\mathbb{E}^n \big|\langle G(v(t), t)\phi, \phi\rangle_{H^\alpha} - \rho_n \langle G^n(t)\phi, \phi\rangle_{H^\alpha}\big| \\ &\leq \big(1 + L\|w\|_{L^2 \times L^2}\big)\|\phi\|_{L^4}^2 \mathbb{E}^n \big\|v(t) - v_t^n\big\|_{L^2} + \big(1 + 2\|f\|_0\big) R(n) \\ &\quad + \delta_+(n) \frac{L\|\phi\|_{L^4}^2}{\pi} \big(\sqrt{|D|}(1 + \|f\|_0)\|\nabla_x w\|_{L^2 \times L^2} + \big\|\nabla_x I(t)\big\|_{L^2}\big) \\ &\quad + \delta_+(n)^2 \frac{1 + 2\|f\|_0}{\pi^2}\|\phi\|_{H^1}^2. \end{aligned}$$

Integrating over $(0, T)$, we obtain with a suitable constant $C_\phi > 0$ independent of n and T the estimate

$$\int_0^T \mathbb{E}^n \left| \left\langle G\big(\nu(t),t\big)\phi,\phi\right\rangle_{H^\alpha} - \rho_n \left\langle G^n(t)\phi,\phi\right\rangle_{H^\alpha} \right| \mathrm{d}t$$
$$\leq C_\phi\Big(\mathbb{E}^n \left\| \nu(t) - \nu_t^n \right\|_{L^1((0,T),L^2)} + TR(n)$$
$$+ T\delta(n)\Big(1 + \left\| \nabla_x I(t) \right\|_{L^1((0,T),L^2)}\Big) + \delta(n)^2\Big).$$

The constant C_ϕ depends on the norm of ϕ in the spaces H 1 (D) H1(D)$H^1(D)$ and $L^4(D)$ where the latter can be estimated in terms of the norm in the Sobolev space $H^\alpha(D)$ due to the embedding $H^\alpha(D) \hookrightarrow L^4(D)$, i.e., C_ϕ is finite and depends only on ϕ∈H α (D) ϕ∈Hα(D)$\phi \in H^\alpha(D)$. Finally, each term in the right-hand side converges to zero for $n \to \infty$, and hence condition (4.33) follows. The proof of Theorem 2.3 is completed.

END NOTE

A normed space X is continuously embedded in another normed space Y, in symbols X↪Y , if X⊂Y and there exists a constant $K < \infty$ such that ‖u‖ Y ≤K‖u‖ X for all u∈X .

The derivation of limit theorems for bounded populations sizes, where $l(k)$ actually is the number of neurons per population, is much more delicate than the subsequent presentation as the transition rate functions become discontinuous. Although this would be a desirable result, we have not yet been able to prove such a theorem, though it is clear that the Wilson–Cowan equation would be the only possible limit. See also a discussion of this aspect in Sect. 3.2.

A continuous embedding of two Hilbert spaces $X \hookrightarrow Y$ is of Hilbert–Schmidt type if for every orthonormal basis ϕ j , of X it holds that $\sum_{j=1}^{\infty} \|\varphi_j\|_Y^2 < \infty$. Then, more precisely, Maurin's theorem states that for non-negative integers m, k, the embedding of $H^{m+k}(D)$ into $H^m(D)$ is of Hilbert–Schmidt type for k>d/2 k>d/2; see [2]. The result was generalised to fractional order Sobolev spaces in [35]: Let D be a bounded, strong local Lipschitz domain in $\mathbb{R}^d$ and $0 \leq \alpha_1 < \alpha_2$ are real numbers. Then it holds that the embedding of $H^{\alpha_2 + d/2}(D)$ into $H^{\alpha_1}(D)$ is of Hilbert–Schmidt type.

Usually, the covariance operator for a Hilbert space-valued process is an operator mapping from the state space into the state space and not into the dual, i.e., in the present situation mapping H −α (D) H−α(D)$H^{-\alpha}(D)$ into itself. Due to the canonical embedding of Hilbert spaces into their dual and the Riesz representation, however, we can effortless change from the usual definition to ours and vice versa. Moreover, the symmetry condition thus implies due to the Hellinger–Toeplitz theorem that the operator is self-adjoint, and hence of trace class if and only if (2.16) is satisfied. The choice of the presentation here is due to the fact that it is simpler to evaluate the duality pairing on $H^{-\alpha}(D)$ than the inner product thereon, as the former usually is just the inner product in $L^2(D)$.

The Poisson process jumps at a faster rate than the components of the Markov chain regardless of the time and the state these are in. Furthermore, all jumps are upward. Hence, using a coupling argument as discussed in the proof of Theorem 4.3.5 in [16], we find that there exists a probability space supporting two processes with distributions equivalent to the Poisson process and the Markov chain component such that the Poisson process dominates the second process for all paths. Clearly, all moments dominate and these inequalities are then valid for any probability spaces supporting these processes.

In the case of an affine function $F(v,t)=f_1(t)\cdot v+f_2(t)$, the mapping f_1 is a linear form on H, which is interchangeable with the expectation operator. For example, in the simplest case, the application of f_1 to $v\in H$ is just a multiplication by a scalar.

ACKNOWLEDGEMENTS

The authors thank J. Touboul for directing our attention also toward the infinite-time convergence in Theorem 2.2. We also thank two anonymous reviewers for their helpful comments.

AUTHORS' CONTRIBUTIONS

MR performed the analysis and wrote and rewrote versions of the manuscript. EB contributed mathematical discussions and to the writing of the manuscript. All authors read and approved the final manuscript.

REFERENCES

1. Acosta G, Durán RG:An optimal Poincaré inequality in L 1 L1L_1 for convex domains. Proc Am Math Soc 2004, 132(1):195–202.

2. Adams RA, Fournier JFJ: Sobolev Spaces. 2nd edition. Academic Press, San Diego; 2003.

3. Benayoun M, Cowan JD, van Drongelen W, Wallace E: Avalanches in a stochastic model of spiking neurons. PLoS Comput Biol 2010., 6(7): Article ID e1000846 Article ID e1000846

4. Blount D: Comparison of stochastic and deterministic models of a linear chemical reaction with diffusion. Ann Probab 1991, 19(4):1440–1462.

5. Bressloff PC: Stochastic neural field theory and the system-size expansion. SIAM J Appl Math 2009, 70: 1488–1521.

6. Bressloff PC: Metastable states and quasicycles in a stochastic Wilson–Cowan model of neuronal population dynamics. Phys Rev E 2010., 82: Article ID 051903 Article ID 051903

7. Bressloff PC: Spatiotemporal dynamics of continuum neural fields. J Phys A, Math Theor 2012., 45: Article ID 033001 Article ID 033001

8. Buice MA, Cowan JD: Field-theoretic approach to fluctuation effects in neural networks. Phys Rev E 2007., 75: Article ID 051919 Article ID 051919

9. Buice MA, Cowan JD, Chow CC: Systematic fluctuation expansion for neural network activity equations. Neural Comput 2010, 22(2):377–426.

10. Corduneanu C: Integral Equations and Applications. Cambridge University Press, Cambridge; 1991.

11. Cowan JD: Stochastic neurodynamics. In Advances in Neural Information Processing Systems. Edited by: Touretzky DS, Lippmann RP, Moody JE. Morgan Kaufman, San Mateo; 1991:62–69.

12. Da Prato G, Zabczyk J: Stochastic Equations in Infinite Dimensions. Cambridge University Press, Cambridge; 1992.

13. Darling RWR, Norris JR: Differential equation approximation for Markov chains. Probab Surv 2007, 5: 37–79.

14. Davis MHA: Markov Models and Optimisation. Chapman & Hall, London; 1993.

15. Ethier SN, Kurtz TG: Markov Processes: Characterization and Convergence. Wiley, New York; 1986.

16. Jacobsen M: Point Process Theory and Applications: Marked Point and Piecewise Deterministic Processes. Birkhäuser, Boston; 2006.

17. Kotelenez P: Law of large numbers and central limit theorem for linear chemical reactions with diffusion. Ann Probab 1986, 14(1):173–193.

18. Kotelenez P: Linear parabolic differential equations as limits of space-time jump Markov processes. J Math Anal Appl 1986, 116(1):42–76.

19. Kurtz TG: Solutions of ordinary differential equations as limits of pure jump Markov processes. J Appl Probab 1970, 7: 49–58.

20. Kurtz TG: Limit theorems for a sequence of jump Markov processes approximating ordinary differential equations. J Appl Probab 1971, 8: 344–356.

21. Mélykúti B, Burrage K, Zygalakis KC: Fast stochastic simulation of biochemical reaction systems by alternative formulations of the chemical Langevin equation. J Chem Phys 2010., 132: Article ID 164109 Article ID 164109

22. Ohira T, Cowan JD: Stochastic neurodynamics and the system size expansion. In Mathematics of Neural Networks: Models, Algorithms, and Applications. Edited by: Ellacott S, Mason JC, Anderson IC. Kluwer Academic, Norwell; 1997:290–294.

23. Payne LE, Weinberger HF: An optimal Poincaré inequality for convex domains. Arch Ration Mech Anal 1960, 5: 286–292.

24. Potthast R, Beim Graben P: Existence and properties of solutions for neural field equations. Math Methods Appl Sci 2010, 33(8):935–949.

25. Prévôt C, Röckner M: A Concise Course on Stochastic Partial Differential Equations. Springer, Berlin; 2007.

26. Riedler MG: Spatio-temporal stochastic hybrid models of excitable biological membranes. PhD thesis. Heriot-Watt University; 2011. Riedler MG: Spatio-temporal stochastic hybrid models of excitable biological membranes. PhD thesis. Heriot-Watt University; 2011.

27. Riedler MG, Thieullen M, Wainrib G: Limit theorems for piecewise deterministic processes in infinite dimensions with applications to diffusion approximation of spatially extended models of neuronal membranes. Electron J Probab 2012., 17: Article ID 55 Article ID 55

28. Sharpe M: General Theory of Markov Processes. Academic Press, San Diego; 1988.

29. Topsoe F: Preservation of weak convergence under mappings. Ann Math Stat 1967, 38(6):1661–1665.

30. Touboul J: Propagation of chaos in neural fields. arXiv:1108.2414v3; 2011. Touboul J: Propagation of chaos in neural fields. arXiv:1108.2414v3; 2011.

31. Touboul J, Ermentrout B: Finite-size and correlation-induced effects in mean-field dynamics. J Comput Neurosci 2011, 31(3):453–484.

32. Touboul J, Hermann G, Faugeras O: Noise-induced behaviors in neural mean field dynamics. SIAM J Appl Dyn Syst 2012, 11(1):49–81.

33. van Kampen NG: Stochastic Processes in Physics and Chemistry. North-Holland, Amsterdam; 1981.

34. Veltz R, Faugeras O: Local/global analysis of the stationary solutions of some neural field equations. SIAM J Appl Dyn Syst 2010, 9: 954–998.

35. Wloka J: Vektorwertige Sobolev-Slobodeckijsche Distributionen. Math Z 1967, 98: 303–318.

36. Yin GG, Zhang Q: Continuous-Time Markov Chains and Applications. Springer, New York; 1998.

Chapter 8

A methodology for stochastic analysis of share prices as Markov chains with finite states

Felix Okoe Mettle[1], Enoch Nii Boi Quaye[1] and Ravenhill Adjetey Laryea[2]

1Department of Statistics, University of Ghana

2Department of Banking and Finance, University of Professional Studies

ABSTRACT

Price volatilities make stock investments risky, leaving investors in critical position when uncertain decision is made. To improve investor evaluation confidence on exchange markets, while not using time series methodology, we specify equity price change as a stochastic process assumed to possess Markov dependency with respective state transition probabilities matrices following the identified state pace (i.e. decrease, stable or increase). We established that identified states communicate, and that the chains are aperiodic and ergodic thus possessing limiting distributions. We developed a methodology for determining expected mean return time for stock price increases and also establish criteria for improving investment decision based on highest transition probabilities, lowest mean return time and highest limiting distributions. We further developed an R algorithm for running the methodology introduced. The established methodology is applied to selected equities from Ghana Stock Exchange weekly trading data.

KEYWORDS

Markov process Transition probability matrix Limiting distribution Expected mean return time Markov chain

1. BACKGROUND

Stock market performance and operation has gained recognition as a significantly viable investment field within financial markets. We most likely find investors seeking to know the background and historical behavior of listed equities to assist investment decision making. Although stock trading is noted for its likelihood of yielding high returns, earnings of market players in part depend on the degree of equity price fluctuations and other market interactions. This makes earnings very volatile, being associated with very high risks and sometimes significant losses.

In stochastic analysis, the Markov chain specifies a system of transitions of an entity from one state to another. Identifying the transition as a random process, the Markov dependency theory emphasizes "memoryless property" i.e. the future state (next step or position) of any process strictly depends on its current state but not its past sequence of experiences noticed over time. Aguilera et al. (1999) noted that daily stock price records do not conform to usual requirements of constant variance assumption in conventional statistical time series. It is indeed noticeable that there may be unusual volatilities, which are unaccounted for due to the assumption of stationary variance in stock prices given past trends. To surmount this problem, models classes specified under the Autoregressive Conditional Heteroskedastic (ARCH) and its Generalized forms (GARCH) make provisions for smoothing unusual volatilities.

Against the characteristics of price fluctuations and randomness which challenges application of some statistical time series models to stock price forecasting, it is explicit that stock price changes over time can be viewed as a stochastic process. Aguilera et al. (1999) and Hassan and Nath (2005) respectively employed Functional Principal Component Analysis (FPCA) and Hidden Markov Model (HMM) to forecast stock price trend based on non-stationary nature of the stochastic processes which generate the same financial prices. Zhang and Zhang (2009) also developed a stochastic stock price forecasting model using Markov chains.

Varied studies (Xi et al.2012; Bulla et al.2010; Ammann and Verhofen2006; and Duffie and Singleton1993) have researched into the application of stochastic probability to portfolio allocation. Building on existing literature, we assume that stock price fluctuations exhibit Markov's dependency and time-homogeneity and we specify a three state Markov process (i.e. price decrease, no change and price increase) and advance the methodology for determining the mean return time for equity price increases and their respective limiting distributions using the generated state-transition matrices. We further replicate the case for a two-state space i.e. decrease in price and increase in price. Based on the methodology, we hypothesize that;

Equity with the highest state transition probability and least mean return time will remain the best choice for an investor.

We explore model performance using weekly historical data from the Ghana Stock Exchange (GSE); we set up the respective transition probability matrix for selected stocks to test the model efficiency and use.

2. REVIEW OF THEORETICAL FRAMEWORK

2.1 Definition of the Markov process

The stochastic process $\{X(t), t \epsilon T\}$ is said to exhibit Markov dependence if for a finite (or countable infinite) set of points $(t_0, t_1, \ldots, t_n, t)$, $t_0 < t_1 < t_2 < \ldots < t_n < t$ where $t, t_r \epsilon T$ $(r = 0, 1, 2, \ldots, n)$.

$$P(X(t) \le x | X(t_n) = x_n, X(t_{n-1}) = x_{n-1}, \ldots, X(t_n) = x_0)$$
$$= P[X(t) \le x | X(t_n) = x_n] = F[X_n, x; t_n, t] \quad (1)$$

From the property given by equation (1), the following relation suffices

$$F(X_n, x; t_n, t) = \int_{y \in S} F(y, x; \tau, t) dF(X_n, y; t_n, \tau) \quad (2)$$

where $t_n < \tau < t$ and S is the state space of the process $\{X(t)\}$.

When the stochastic process has discrete state and parameter space, (2) takes the following form: for $n > n_1 > n_2 > \ldots > n_k$ and $n, n_r \epsilon T$ $(r = 1, 2, \ldots, k)$

$$P(X_n = j | X_{n_1} = i_1, X_{n_2} = i_2, \ldots, X_{n_k} = i_k)$$
$$= P\Big(X_n = j | X_{n_1} = i_1) = P_{ij}^{(n_k, n)} \quad (3)$$

A stochastic process with discrete state and parameter spaces which exhibits Markov dependency as in (3) is known as a Markov Process.

From the Markov property, for $n_k < r < n$ we get

$$\begin{aligned} P_{ij}^{(n_k,n)} &= P(X_n = j | X_{n_k} = i) \\ &= \sum_{m \in S} P(X_n = j | X_r = m) P(X_r = m | X_{n_k} = i) \\ &= \sum_{m \in S} P_{ij}^{(n_k,r)} P_{mj}^{(r,n)} \end{aligned} \tag{4}$$

equations (2) and (4) are known as the Chapman-Kolmogorov equations for the process.

2.2 n-step transition probability matrix and n-step transition probabilities

If P is the transition probability matrix of a Markov chain $\{X_n, n = 0, 1, 2, \ldots\}$ with state space S, then the elements of P^n (*P raised to the power* $P_{ij}^{(n)}\ i, j \epsilon S$ are the n-step transition probabilities where $P_{ij}^{(n)}$ is the probability that the process will be in state j at the nth step starting from state i.

The above statement can clearly be shown from the Chapman-Kolmogorov equation (4) as follows; for a given r and s, write

$$P_{ij}^{(s+r)} = \sum_{k \in s} P_{ik}^{(r)} P_{kj}^{(S)}$$

Set $r = 1$, $s = 1$ in the above equation to get

$$P_{ij}^{(2)} = \sum_{k \in s} P_{jk} P_{kj}$$

Clearly, $P_{ij}^{(2)}$ is the $(i, j)th$ element for the matrix product $P \times P = P^2$. Now suppose $P_{ij}^{(r)}$ ($r = 3, 4, \ldots, n$) is the $(i, j)^{th}$ of P^r then by the Kolmogorov equation, the

$$P_{ij}^{(r+1)} = \sum_{k \in S} P_{ik}^{(r)} P_{kj}$$

which again can be seen as the $(i, j)^{th}$ element of the matrix product $P^r P = P^{r+1}$. Hence by induction, $P_{ij}^{(n)}$ is the $(i, j)^{th}$ element of P^n $n = 2, 3, \ldots$.

To specify the model, the underlying assumption is stated about the identified n-step transition probability (stating without proof).

The transition probability matrix is accessible with existing state communication. Further, there exists recurrence and transience of states. States are also assumed to be irreducible and belong to one class with the same period which we take on the value 1. Thus the states are aperiodic.

2.3 Limiting distribution of a Markov chain

If P is the transition probability matrix of an aperiodic, irreducible, finite state Markov chain, then

$$\lim_{t\to\infty} P^t = \pi = \begin{bmatrix} \alpha \\ \alpha \\ \vdots \\ \alpha \end{bmatrix} \tag{5}$$

Where $\alpha = [\alpha_1, \alpha_2, \ldots, \alpha_m]$ with $0 < \alpha_j < 1$ and $\sum_{j=1}^{m} \alpha_j = 1.$ See Bhat (1984). The chain with this property is said to be ergodic and has a limiting distribution $\boldsymbol{\pi}$. The transition probability matrix P of such a chain is primitive.

2.4 Recurrence and transience of state

Let X_t be a Markov Chain with state space S, then the probability of the first transition to state j at the tth step starting from state i is

$$f_{ij}^{(t)} = P[X_t = j,\ X_r \neq j;\ r = 1, 2, 3, \ldots, t-1 | X_0 = i] \tag{6}$$

Thus the probability that the chain ever returns to state j is

$$f_{ij} = \sum_{t=1}^{\infty} f_{ij}(t)$$

and μ $\mu_{ij} = \sum_{t=1}^{\infty} t f_{ij}^{(t)}$ is the expected value of first passage time. Further, if $i = j$, then;

$$f_{ii}^{(t)} = P[X_t = i,\ X_r \neq i;\ r = 1, 2, 3, \ldots, t-1 | X_0 = i] \tag{7}$$

$\mu_{ii} = \mu_i = \sum_{t=1}^{\infty} t f_{ii}^{(t)}$ is the mean recurrence time of state i if state i is recurrent.

A state i is said to be recurrent (persistent) if and only if, starting from state i, eventual return to this state is certain. Thus state i is recurrent if and only if

$$f_{ii}^{*} = \sum_{t=1}^{\infty} f_{ii}^{(t)} = 1 \tag{8}$$

A state i is said to be transient if and only if, starting from state i, there is a positive probability that the process may not eventually return to this state. This means $f_{ii}^{*} < 1$

3. MODEL SPECIFICATION

3.1 Defining the problem (Equity price changes as a three-state Markov process)

Let Y_t be the equity price at time t where $t = 0, 1, 2, \ldots, n$ (t is measured in weekly time intervals). Further, we define $d_t = Y_t - Y_{t-1}$ which measures the change in equity price at time t. Considering each closing week's price as discrete time unit for which we define a random variable X_t to indicate the state of equity closing price at time t, a vector spanned by 0, 1, 2

$$X_t = \begin{cases} 0 & if \quad d_t < 0 \quad decrease\, in\, equity\, price\, from\; t-1\; to\, t \\ 1 & if \quad d_t = 0 \quad no\, change\, in\, equity\, price\, from\; time\, t-1\; to\, t \\ 2 & if \quad d_t > 0 \quad increase\, in\, equity\, price\, from\; time\, t-1\; to\, t \end{cases}$$

Next, we define an indicator vector

$$I_{i,t} = \begin{cases} 1 & if \quad X_t = i \\ 0 & if \quad X_t \neq i \end{cases} \; for\; i = 0,\, 1,\, 2 \;\; and \;\; t = 1,\, 2,\, \ldots,\, n \tag{9}$$

Then clearly for the outcome of X_t we have

$$n_i = \sum_{t=1}^{n} I_{i,t} \; for \quad i = 0,\, 1,\, 2 \tag{10}$$

where $n = \sum_{i=0}^{2} n_i$. Hence estimates of the probability that the equity price reduce, did not change and increased can be obtained respectively by

$$\hat{P}_0 = \frac{n_0}{n}, \quad \hat{P}_1 = \frac{n_1}{n} \quad and \quad \hat{P}_2 = \frac{n_2}{n} \tag{11}$$

For the stochastic process X_t obtained above for $t = 1, 2, \ldots, n$ we can obtained estimates of the transition probabilities $P_{ij} = \Pr(X_t = j | X_{t-1} = i)$ for $j = 0, 1, 2$ by defining

$$\delta_t^{(ij)} = \begin{cases} 1 & if\ X_t = i\ and\ X_{t-1} = j \\ 0 & otherwise \end{cases} \quad \begin{matrix} for\ t = 1,\ 2,\ \ldots,\ n-1 \\ and\ i,j = 0,\ 1,\ \ldots, k \end{matrix}$$

where $k + 1$ is the number of states of the chain.

$$n_{ij} = \sum_{t=1}^{n-1} \delta_t^{(i,j)} for \quad i,j = 0,1,2. \quad Then \quad \hat{P}_{ij} = \frac{n_{ij}}{n_i} \\ for \quad i,j = 0,1,\ldots,k \tag{12a}$$

Therefore, an estimate for the transition matrix for $k = 2$ is

$$\hat{\boldsymbol{P}} = \begin{bmatrix} \hat{P}_{00} & \hat{P}_{01} & \hat{P}_{02} \\ \hat{P}_{10} & \hat{P}_{11} & \hat{P}_{12} \\ \hat{P}_{20} & \hat{P}_{21} & \hat{P}_{22} \end{bmatrix} \tag{12b}$$

Suppose the data in Additional file1 is uploaded as .csv, then *R* code for computing estimates in (12b) can be found in Additional file2 (three-state Markov Chain function column).

```
Y=read.csv(#the data can be obtained in the additional file)

R<-function(D){Y=as.numeric(D)
                DIFF <- Y
                for(i in 1:length(DIFF)){
                X<-function(DIFF){if (DIFF<0) X=0
                else if (DIFF==0) X=1 else X=2
 }

 x<-c(lapply(DIFF,X))
}

X<-as.numeric(x)
        t=seq(1:length(X))
        I<-function(i){ ifelse(X[1:(length(X))] == i,1,0)
}

n<-function(i){sum(I(i),na.rm=T)}

n1=c(n(0),n(1),n(2))
N=sum(n1)
p<-function(i){n(i)/N}
t=seq(1:length(X))

Z<- function(i,j) {return(ifelse(X[t]==i&X[t+1]==j,1,0))
}

P<-function(i,j){sum(Z(i,j),na.rm=T)/n(i)
}

for(i in 0:2){
  for(j in 0:2){
   P(i,j)}
  TP<-matrix(c(P(0,0),P(1,0),P(2,0),P(0,1),P(1,1),P(2,1),P(0,2),P(1,2),P(2,2)),3,3)
  colnames(TP)=c("D","S","I");rownames(TP)=c("D","S","I")
```

3.2 For a two-state Markov process

We maintain the above defined terms and set

$$X_t = \begin{cases} 0 & if \quad d_t \leq 0 \quad no\ increase\ in\ equity\ price\ from\ \ t-1\ \ to\ \ t \\ 1 & if \quad d_t > 0 \quad increase\ in\ equity\ price\ from\ time\ \ t-1\ \ to\ \ t \end{cases}$$

further set $i, j = 0, 1$, (for $k = 1$) and apply (9), (10), (11), (12a), and (12b) sequentially, we obtain

$$\hat{P} = \begin{bmatrix} \hat{P}_{00} & \hat{P}_{01} \\ \hat{P}_{10} & \hat{P}_{11} \end{bmatrix}$$

without loss of generality, suppose X_t has state space $s = \{0, 1\}$ and transition probability matrix

$$\boldsymbol{P} = \begin{pmatrix} 1-\theta & \theta \\ \beta & 1-\beta \end{pmatrix}, \quad 0 < \alpha, \beta < 1 \tag{13}$$

Then, $f_{00}^{(1)} = 1 - \vartheta$ and for $n \geq 2$, we have;

$$\begin{aligned} f_{00}^{(t)} &= P[X_t = 0; X_r \neq 0; r = 1,\ 2,\ 3,\ \ldots, t-1 | X_0 = 0] \\ &= P[X_t = 0; X_r = 1; r = 1,\ 2,\ 3,\ \ldots,\ t-1 | X_0 = 0] \end{aligned}$$

By the Markov property and the definition of conditional probability, we have

$$\begin{aligned} f_{00}^{(n)} &= P\left[X_t = 0 | X_{t-1} = 1\right] \left[\prod_{r=2}^{t-1} P[X_r = 1 | X_{r-1} = 1]\right] P[X_1 = 1 | X_0 = 0] \\ &= \beta(1-\beta)^{t-2}\theta = \theta\beta(1-\beta)^{t-2} \quad t \geq 2 \end{aligned} \tag{14}$$

solving $\mu_0 = \mu_{00} = \sum_{t=1}^{\infty} t f_{00}^{(t)}$ to obtain the respective mean recurrence time. Thus,

$$\begin{aligned} \mu_0 = \mu_{00} = \sum_{t=1}^{\infty} t f_{00}^{(t)} &= 1-\theta + \sum_{t=2}^{\infty} t\theta\beta(1-\beta)^{t-2} \\ &= \left(\frac{\theta+\beta}{\beta}\right) \end{aligned} \tag{15}$$

Similarly, we have

$$\begin{array}{lll} f_{01}^{(t)} = \theta(1-\theta)^{t-1}, & t \geq 1 & \mu_{01} = 1 \\ f_{10}^{(t)} = \beta(1-\beta)^{t-1}, & t \geq 1 & \mu_{10} = 1 \\ f_{11}^{(1)} = (1-\beta), & t = 1 & \\ f_{11}^{(t)} = \theta\beta(1-\alpha)^{t-2}, & t \geq 2 & \mu_{11} = \mu_1 = \left(\dfrac{\theta+\beta}{\theta}\right) \end{array} \tag{16}$$

With the corresponding R algorithm shown in Additional file2 (two-state Markov Chain function column).

```
DIFF <- Y[-1]-Y[1:(length(Y)-1)]

for(i in 1:length(DIFF)){X<-function(DIFF){
          if (DIFF<=0) X=0
          else if (DIFF>0) X=1  }
 x<-c(lapply(DIFF,X))
}

X<-as.numeric(x)

t=seq(1:length(X))
 I<-function(i){ifelse(X[t] == i,1,0)
 }

 n<-function(i){sum(I(i),na.rm=T)}
 n1=sum(c(n(0),n(1))
 p<-function(i){n(i)/n1}
 t=seq(1:length(X))

 N<- function(i,j) {return(ifelse(X[t]==i&X[t+1]==j,1,0))
 }

 P<-function(i,j){sum(N(i,j),na.rm=T)/n(i)
 }

for(i in 0:1){
for(j in 0:1){
P(i,j)}
TP<-matrix(c(P(0,0),P(1,0),P(0,1),P(1,1)),2,2)
colnames(TP)=c("DS","I");rownames(TP)=c("DS","I")
 }

 i=c(0:2)
 j=c(0:2)
 s<-function(b){
  alpha=0.001
  c=1
  r=(1-2*min(TP[i,j]))
  nthlimit=(log10(alpha)-log10(c))/log10(r)
  nthlimit
 }
 s1<-function(b){TP1=TP%*%TP
  alpha=0.001
  c=(1-2*min(TP1[i,j]))^(-1)
  r=(1-2*min(TP1[i,j]))^1/2
  nthlimit=(log10(alpha)-log10(c))/(log10(r))
  nthlimit
 }
 regular=all(c(TP[i,j])>0)

 nth<-function(){ ifelse(regular,s(b),s1(b))
 }
```

3.3 Generating eigen vectors for computation of limiting distributions

After the transition probabilities are obtained for both two-state and three-state chains, the R codes in the lower portions of columns one and two in Additional file2 were used to generate the respective eigen vectors for computation of limiting distributions.

```
A=eigen(t(TP))
A=Re(A$vec[,1])
STP=A/sum(A)

i=c(1:3)
j=c(1:3)
 s<-function(b){alpha=0.1
  c=1
  r=(1-2*min(TP[i,j]))
  nthlimit=(log10(alpha)-log10(c))/log10(r)
  nthlimit

 }
 s1<-function(b){TP1=TP%*%TP
  alpha=0.1
  c=(1-2*min(TP1[i,j]))^(-1)
  r=(1-2*min(TP1[i,j]))^1/2
  nthlimit=(log10(alpha)-log10(c))/(log10(r))
  nthlimit
 }

 nth<-function(){ifelse(all(c(TP[i,j])>0),s(b),s1(b)+3)
 }

Power <- function(P,n){
A <-diag(rep(1,dim(P)[1]))
for(i in 1:n)
{A <- A %*% P}
return(A)
}
K=Power(TP,nth())
rownames(K)=c("D","S","I")

NUMBSTATES=c(n(0),n(1),n(2))
nthweek=nth()
PROBABILITIES=c(p(0),p(1),p(2))
options(digits=2)
TRANSPROB=TP
STATIONDIST= STP
LIMITPROB=K
return(list(NUMBSTATES=c(n(0),n(1),n(2)),PROBABILITIES=c(p(0),p(1),p(2)),nthweek=nt
h(),TRANSPROB=TP,STATIONDIST=STP, LIMITPROB=K))
}
```

4. FINDINGS AND DISCUSSIONS

4.1 Data structure and summary statistics

Data used for this paper are weekly trading price changes for five randomly selected equities on the Ghana Stock Exchange (GSE), each covering period starting from January 2012-December 2013. We obtain the weekly price changes using the relation $d_t = Y_t - Y_{t-1}$ where Y_t represents the equity closing price on week t and Y_{t-1} is the opening price for the immediate past week. The equities selected include Aluworks (ALW), Cal Bank (CAL), Ecobank Ghana (EBG), Ecobank Transnational Incorporated (ETI), and Fan Milk Ghana Limited (FML).

In all, 104 (52 weeks) observational data points where obtained. Summary statistics on all respective equities on the GSE are shown in Table 1. We present summaries on the respective number of weekly price decreases, no change in price and price increase. Descriptive statistics for each equity weekly price change is also shown.

Overall, the frequency of "no price change" was more experienced over the study period. The lowest and highest price changes for the trading period are respectively -4.19 and 9.54. The estimated values of the kurtosis and skewness are also shown. Figure 1 presents a plot of the average weekly equity price changes of respective equities listed on the GSE over the study period in comparison to the standard deviation of weekly price changes.

Table 1. Summary statistics on the weekly trading price change over the study period

	Number of weekly price change			Weekly price change summary						
	Decrease	No change	Increase	Mean	SD	Max	Min	Skew.	Kurt.	Count
ALW	15	77	12	0.00	0.01	0.01	-0.04	-2.30	14.23	104
AYRIN	8	89	7	0.00	0.00	0.01	-0.01	-0.10	4.39	104
BOPP	26	45	33	0.01	0.13	0.44	-0.62	-1.80	11.79	104
CAL	27	40	37	0.00	0.03	0.12	-0.07	1.69	7.26	104
EBG	30	44	30	0.00	0.19	0.50	-1.60	-5.65	51.32	104
EGL	21	46	37	0.01	0.06	0.39	-0.25	1.05	15.79	104
ETI	18	59	27	0.00	0.01	0.04	0.04	-0.71	5.62	104
FML	22	38	44	0.03	0.12	0.85	0.19	4.08	25.22	104
GCB	25	37	42	0.02	0.13	0.79	-0.41	1.68	12.51	104
GGBL	5	51	48	0.04	0.10	0.73	-0.20	3.84	22.70	104
GLD	7	79	18	0.04	0.34	3.13	-0.72	7.22	64.43	104
GOIL	16	53	35	0.00	0.03	0.12	-0.23	-2.99	23.63	104
HFC	8	75	21	0.01	0.03	0.27	-0.08	5.76	47.80	104
MLC	7	75	22	0.00	0.01	0.05	-0.03	1.04	5.42	104
PBC	13	81	10	0.00	0.01	0.04	-0.02	1.53	11.02	104
PZC	22	55	27	0.01	0.38	3.02	-1.00	4.78	39.15	104
SCB	38	40	26	0.14	1.30	9.54	4.19	4.45	30.38	104

SCBPREF	11	87	6	0.00	0.01	0.01	-0.03	-2.35	9.67	104
SIC	19	67	18	0.00	0.02	0.16	-0.06	5.55	50.15	104
SOGEGH	12	89	3	0.00	0.02	0.01	-0.18	6.56	50.60	104
SWL	9	84	11	0.01	0.45	3.15	-2.00	2.28	27.27	104
TBL	21	62	21	0.02	0.62	2.99	-3.00	0.16	12.33	104
TLW	16	56	32	0.23	0.97	6.56	-1.97	3.77	19.38	104
TOTAL	16	66	22	0.01	0.08	0.52	-0.16	4.89	29.16	104
TRANSOL	12	63	29	0.04	0.18	1.26	-0.50	4.24	24.87	104
UNIL	3	76	25	0.03	0.23	1.79	-0.77	5.23	38.66	104
UTB	12	87	5	0.00	0.01	0.02	-0.02	-1.30	6.18	104

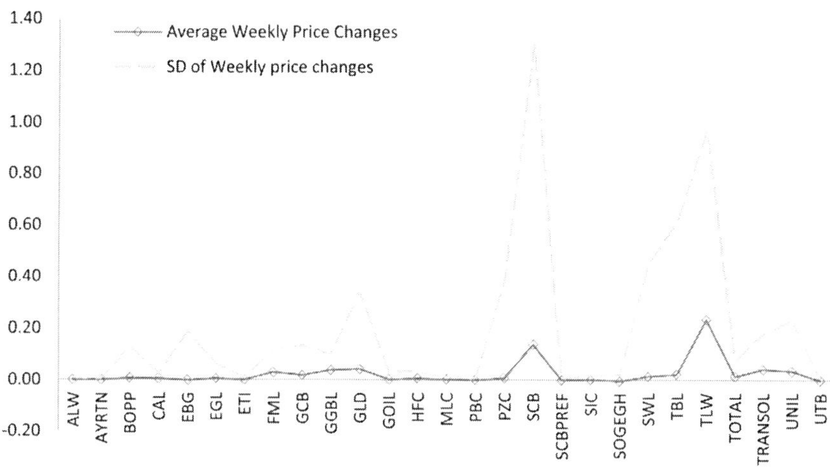

Figure 1. A plot of mean and standard deviation of weekly price changes of equities. The plot indicates a very volatile weekly market price fluctuation for any market participating investor. This indicates high level of risk associated with equity purchase decision. We consider that the rational investor would basically seek to maximize purchasing decisions faced with this risk.

4.2 Empirical results on model application (three-state Markov chain)

For the five randomly selected equities, the transition probabilities of the equities are presented as follows. These were obtained from equation (12a) defining $P_{ij} = \frac{n_{ij}}{n_i}$ w.r.t. the three-state space Markov process. A 3 × 3 transition matrix is obtained for respective equities as defined by (12b).

From the results of the algorithm, we select 5 equities with which we implement the hypothesis. They include;

ALW transition probability matrix

$$\hat{P} = \begin{bmatrix} 0.133333 & 0.666667 & 0.200000 \\ 0.139241 & 0.759494 & 0.101266 \\ 0.166667 & 0.750000 & 0.083333 \end{bmatrix}$$

CAL transition probability matrix

$$\hat{P} = \begin{bmatrix} 0.296296 & 0.407407 & 0.296296 \\ 0.261905 & 0.476190 & 0.261905 \\ 0.189189 & 0.324324 & 0.486486 \end{bmatrix}$$

EBG transition probability matrix

$$\hat{P} = \begin{bmatrix} 0.433333 & 0.366667 & 0.200000 \\ 0.255319 & 0.553191 & 0.191489 \\ 0.137931 & 0.344828 & 0.517241 \end{bmatrix}$$

ETI transition probability matrix

$$\hat{P} = \begin{bmatrix} 0.166667 & 0.611111 & 0.222222 \\ 0.131148 & 0.639344 & 0.229508 \\ 0.259259 & 0.444444 & 0.296296 \end{bmatrix}$$

FML transition probability matrix

$$\hat{P} = \begin{bmatrix} 0.380952 & 0.523810 & 0.095238 \\ 0.170732 & 0.487805 & 0.341463 \\ 0.136364 & 0.227273 & 0.636364 \end{bmatrix}$$

Clearly, $\hat{P}_{ij} > 0$ for all $i, j = 0, 1, 2$ indicating irreducibility of the chains for all equities. Hence state 0 for all the equities is aperiodic and since periodicity is a class property, the chains are aperiodic. These imply that the chains are ergodic and have limiting distributions.

Figure 2 presents the *t-step* transition probabilities for share price increases based on the assumption of time-homogeneity. This shows linear plot of transition probabilities for $P_{22}^{(t)}$ for each selected stock as computed above. It measures the probability that a share at initial state (*i. e. state* 2) at inception transited to *state* 2 again after *t weeks*. Regarding the plot of the transition probabilities, the logical reasoning is to choose the equity which has the highest P_{22}.

From the plot, *FML* share is the best choice for the investor since the probability that it increases from a high price to another higher price is higher when compared to the other selected stocks. ALW recorded the least probability of transition within the period. Comparing CAL to EBG, the methodology shows that CAL shares maintain high probability of moving to higher prices as compared to EBG shares although the later started with high prices at inception.

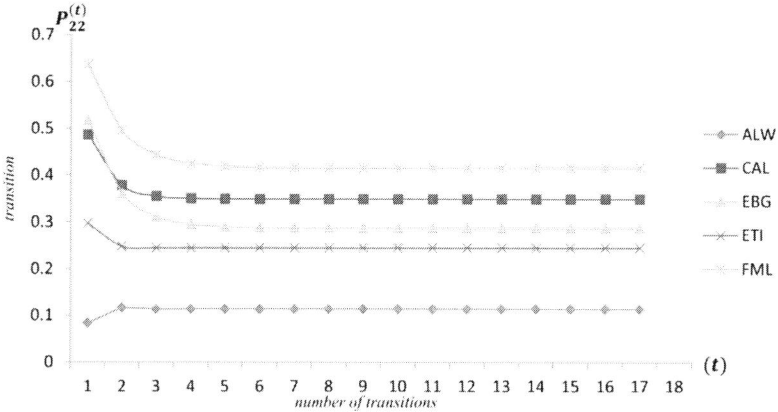

Figure 2. t-step transition probabilities for share price increases.

Using equation (5), the limiting distributions of the respective equities were computed. These probabilities measure the proportions of times the equity states within a particular state in the long run. From Table 2, ALW equity has 14% chance of reducing and 11% chance of increasing in the long run. It however has 75% chance of no change in price. Similarly, in the long run, FML equity has 20% chance of reducing, 39% chance of experiencing no change in price and 42% chance of increasing in price. It is easily seen that for this instance, FML equity has the highest probability of price increase in the long run.

Table 2. Entries of the limiting distribution at for respective equities

Equity	Limiting distribution		
	a_1	a_2	a_3
ALW	0.141509	0.745283	0.113208
CAL	0.244980	0.406396	0.348625
EBG	0.269568	0.443025	0.287407
ETI	0.168470	0.586912	0.244618
FML	0.198113	0.386792	0.415094

4.3 Empirical model application (the two-state Markov process)

Defining a two-state space Markov process following from equation (13), we derive the state transition probabilities. The two-state transition probability matrix entries are indicated in Table 3 below;

Table 3. Entries of two-state transition matrices for selected equities

Equities	P_{00}	P_{01}	P_{10}	P_{11}
	$1-\theta$	θ	β	$1-\beta$
ALW	0.133333	0.866667	0.142857	0.857143
CAL	0.296296	0.703704	0.227848	0.772152
EBG	0.433333	0.566667	0.210526	0.789474
ETI	0.166667	0.833333	0.170455	0.829545
FML	0.380952	0.619048	0.152941	0.847059

Applying equations (15) and (16) to the transition probabilities, we obtain the respective mean return time of the selected equities. These are shown in Table 4 below;

Table 4. Expected mean return time for respective stocks

Equity	μ_{00}	μ_{11}
	$\frac{(\theta+\beta)}{\beta}$	$\frac{\theta+\beta}{\theta}$
ALW	1.1555556	7.4285714
CAL	1.3837280	3.6060127
EBG	1.5488889	2.8218623
ETI	1.2009132	5.9772727
FML	1.4497355	3.2235294

Mean return time is measured in weeks with μ_{ij} as defined in (15) and (16). The mean return time measures the expected time until the equity price's next return to the state it was initially in at time 0. Figure 3 presents a plot of expected return time of the selected stocks at μ_{11}. This determines the expected time until the next increase in share. We expect that the choice of share should not only have the highest transition probability, but should relatively possess a lower mean return time. Possessing the least mean return time for μ_{11} signifies the shortest return time to a price increase.

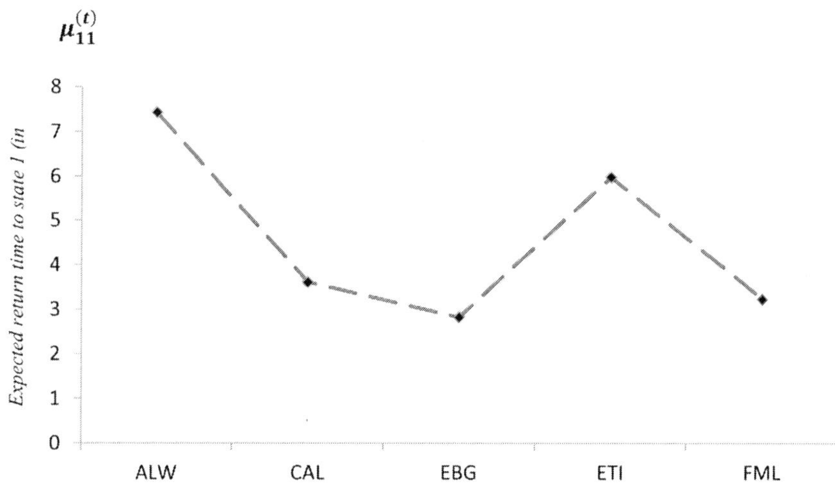

Figure 3. Mean recurrence time of selected shares.

5. CONCLUSION

The Markov Process provides a credible approach for successfully analyzing and predicting time series data which reflect Markov dependency. The study finds that all states obtained communicate and are aperiodic and ergodic hence possessing limiting distributions. It is distinctive from Figures 1 and2 (expected return time and t-step state transition probabilities of equity price increases i.e. P_{ij} transition from state 2 to state 2) that the investor gains good knowledge about the characteristics of the respective equities hence improving decision making in the light return maximization. With regards to the selected stocks, FML equity recorded the highest state transition probabilities, highest limiting distribution but the second lowest mean return time to price increases (i.e. 3.224 weeks).

Our suggested use of Markov chains as a tool for improving stock trading decisions indeed aids in improving investor knowledge and chances of higher returns given risk minimization through best choice decision. We showed that the proposed method of using Markov chains as a stochastic analysis method in equity price studies truly improves equity portfolio decisions with strong statistical foundation. In our future work, we shall explore the case of specifying an infinite state space for the Markov chains model in stock investment decision making.

AUTHORS' CONTRIBUTIONS

FOM introduced the idea, undertook the theoretical and methodology development. ENBQ developed the codes and helped in the analysis and typesetting of mathematical equations. RAL also helped in the analysis and the general typesetting of the manuscript. All authors read and approved the final manuscript.

REFERENCES

1. Aguilera MA, Ocaña AF, Valderrama JM Applied stochastic models in business and industry. In Stochastic modeling for evolution of stock prices by means of functional principal component analysis. John Wiley & Sons, Ltd., New York; 1999.

2. Ammann M, Verhofen M Working paper series in Finance. No. 20. The effect of market regimes on style allocation 2006.*http://www.finance.unisg.ch*

3. Bhat UN Wiley series in Probability & Mathematical Statistics. Elements of applied stochastic processes 2nd edition. 1984.

4. Bulla J, Mergner S, Bulla I, Sesboüé A, Chesneau C Munich Personal RePEc Archive. Markov-switching asset allocation: do profitable strategies exist? 2010. MPRA Paper no. 21154.*http://mpra.ub.uni-muenchen.de/21154/*

5. Duffie D, Singleton JK: Simulated moments estimation of Markov models of asset prices. Econometrica 1993, 61(4):929-952. 10.2307/2951768

6. Hassan RM, Nath B Proceedings of the 2005 5th international conference on intelligent systems design and applications (ISDA'05). In Stock market forecasting using hidden Markov model: a new approach. IEEE Computer Society, Washington, DC, USA; 2005.

7. Xi X, Mamon R, Davison M: A higher-order hidden Markov chain-modulated model for asset allocation. J Math Model Algorithm Oper Res 2012, 13(1):59-85.

8. Zhang D, Zhang X: Study on forecasting the stock market trend based on stochastic analysis method. Int J Bus Manage 2009, 4(6):163-170.

Chapter 9

The Analysis of Real Data Using a Stochastic Dynamical System Able to Model Spiky Prices

Lorella Fatone[1], Francesca Mariani[2], Maria Cristina Recchioni[3], Francesco Zirilli[4]

[1]Dipartimento di Matematica e Informatica, Università di Camerino, Camerino, Italy

[2]CERI—Centro di Ricerca "Previsione, Prevenzione e Controllo dei Rischi Geologici", Università di Roma "La Sapienza", Roma, Italy

[3]Dipartimento di Scienze Sociali "D. Serrani", Università Politecnica delle Marche, Ancona, Italy

[4]Dipartimento di Matematica "G. Castelnuovo", Università di Roma "La Sapienza", Roma, Italy

ABSTRACT

In this paper we use filtering and maximum likelihood methods to solve a calibration problem for a stochastic dynamical system used to model spiky asset prices. The data used in the calibration problem are the observations at discrete times of the asset price. The model considered has been introduced by V. A. Kholodnyi in [1,2] and describes spiky asset prices as the product of two independent stochastic processes: the spike process and the process that represents the asset prices in absence of spikes. A Markov chain is used to regulate the transitions between presence and absence of spikes. As suggested in [3] in a different context the calibration problem for this model is translated in a maximum likelihood problem with the likelihood function defined through the solution of a filtering problem. The estimated values of the model parameters are the coordinates of a constrained maximizer of the likelihood function. Furthermore, given the calibrated model, we develop a sort of tracking procedure able to forecast forward asset prices. Numerical examples using synthetic and real data of the solution of the calibration problem and of the performance of the tracking procedure are presented. The real data studied are electric power price data taken from the UK electricity market in the years 2004-2009. After calibrating the model using the spot prices, the forward prices forecasted with the tracking procedure and the observed forward prices are compared. This comparison can be seen as a way to validate the model, the formulation and the solution of the calibration problem and the forecasting procedure. The result of the comparison is satisfactory.

In the website:http://www.econ.univpm.it/recchioni/finance/w10 some auxiliary material including animations that helps the understanding of this paper is shown. A more general reference to the work of the authors and of their coauthors in mathematical finance is the website: http://www.econ.univpm.it/recchioni/finance.

KEYWORDS

stochastic dynamical system; inverse problem; spiky prices; data time series analysis

1. INTRODUCTION

A spike in an asset price is an abrupt movement of the price followed by an abrupt movement of approximately the same magnitude in the opposite direction. The modeling of spikes in asset prices is a key problem in finance. In fact spiky prices are encountered in several contexts such as, for example, in electric power prices and, more in general, in commodity prices.

In this paper the model introduced by V. A. Kholodnyi in [1,2] to describe spiky prices is combined with some ideas introduced by the authors and some coauthors to study calibration problems in mathematical finance, see [3-9]. That is we introduce a stochastic dynamical system to model spikes in asset prices and we study a calibration problem for the dynamical system introduced. The method proposed to solve the calibration problem is tested doing the analysis of data time series. We consider synthetic and real data. The real data studied are electric power price data taken from the UK electricity market in the years 2004-2009.

Following V. A. Kholodnyi (see [1,2]) we model spiky asset prices as a stochastic process that can be represented as the product of two independent stochastic processes: the spike process and the process that describes the asset prices in absence of spikes. The spike process models spikes in asset prices and it is either equal to the multiplicative amplitude of the spike during the spike periods or equal to one during the regular periods, that is during the periods between spikes. The second stochastic process of the Kholodnyi model describes prices in absence of spikes. This last process has been chosen as a diffusion process. Finally we use a two-state Markov chain in continuous time to determine whether asset prices are in the spike state, that is during a spike, or in the regular state, that is between spikes.

The model for spiky asset prices studied depends on five real parameters. Two of them come from the process that describes the asset prices in absence of spikes, one of them comes from the spike process and the last two parameters come from the two-state Markov chain used to model the transitions between spike and regular states.

The calibration problem studied consists in estimating these five parameters from the knowledge at discrete times of the asset prices (observations of the spiky prices). That is the calibration problem is a parameter identification problem or, more in general, is an inverse problem for the stochastic dynamical system that models the asset prices. This calibration problem is translated in a constrained optimization problem for a likelihood function (maximum likelihood problem) with the likelihood function defined through the solution of a filtering problem. The likelihood function is defined using the probability density function associated with the diffusion process modeling asset prices in absence of spikes. This formulation of the calibration problem is inspired to the one introduced in [3] in the study of the Heston stochastic volatility model that has been later extended to the study of several other calibration problems in mathematical finance (see [4,5,8, 9]).

The filtering and the maximum likelihood problems mentioned above are solved numerically. The resulting numerical solution of the calibration problem determines the values of the (unknown) parameters that make most likely the observations actually made. Note that in the processing of numerical data to improve the robustness and the quality of the solution of the calibration problem some preliminary steps are introduced in the optimization procedure used to solve the calibration problem and the results obtained in these preliminary steps are used to penalize the likelihood function obtained from the filtering problem. That is the maximum likelihood problem originally formulated in analogy to [3] is reformulated adding penalization terms to the likelihood function and choosing an ad hoc initial guess for the optimization procedure to improve the robustness and the quality of its solution. This reformulated optimization problem is solved numerically using a method based on a variable metric steepest ascent method.

Furthermore, as a byproduct of the solution of the filtering problem, we develop a tracking procedure that, given the calibrated model, is able to forecast forward asset prices.

The method used to solve the calibration problem and the tracking procedure are used to analyze data time series. Numerical examples of the solution of the calibration problem and of the performance of the tracking procedure using synthetic and real data are presented. The synthetic data are obtained computing one trajectory of the stochastic dynamical system that models spiky asset prices. We generate daily synthetic data for a period of two years. The first year of data is generated with one choice of the model parameters, the second year of data is generated with a different choice of the model parameters. The second year of data is generated using as initial point of the trajectory the last point of the first year of data. In the solution of the calibration problem we choose as observation period a period of one year, that is we use as data the daily observations corresponding to a time period of one year, and we move one day at the time this observation period through the two years of data. The calibration problem is solved for each choice of the observation period. The two choices of the model parameters used to generate the data and the time when the model parameters change value are reconstructed satisfactorily by the calibration procedure. The real data studied are daily electric power price data taken from the UK. electricity market. These electric power price data are spiky data. We choose the data of the calibration problems considered as done above in the study of synthetic data extracting the observation periods from a time series of five years (i.e. the years 2004-2009) of daily electric power (spot) price data taken from the UK market. The results obtained show that the model is able to establish a stable relationship between the data time series and the estimated model parameter values. Note that in the real data time series for each observation day we have the electric power spot price and the associated forward prices observed that day for a variety of delivery periods. That is for each spot price there is a set of forward prices associated to it corresponding to different delivery periods. Moreover in the calibration problem only spot prices are used as data. To exploit this fact we proceed as follows. After calibrating the model using as data the spot prices observed in the first three years of the data time series, we use the calibrated model, the tracking procedure and the spot prices not used in the calibration to forecast the forward prices associated to these last spot prices. We compare the forward prices forecasted with the tracking procedure with the observed forward prices. The comparison is satisfactory and establishes the effectiveness of the model, the validity of the proposed formulation and solution of the calibration problem and the quality of the forecasted prices.

We note that the model studied is too simplistic to be of practical value in electricity markets. In fact our model is able to capture only one property of the electric power prices: the presence of spikes. It does not consider, for example, the mean-reverting property and the presence of weekly and season cycles in electricity prices. This study aims to be a first attempt to solve, with the strategy presented in Section 3, calibration problems involving stochastic dynamical systems that can be used to describe electric power prices. That is the methodology discussed in this paper can be applied in the calibration of more sophisticated stochastic dynamical models that can be used in electricity markets (see for example [10]).

The website: http://www.econ.univpm.it/recchioni/finance/w10 contains some auxiliary material including some animations that helps the understanding of this paper. A general reference to the work of the authors and of their coauthors in mathematical finance is the website: http://www.econ.univpm.it/recchioni/finance.

The paper is organized as follows. In Section 2 we describe the model for spiky asset prices. In Section 3 we formulate and we solve the filtering and the calibration problems for the model presented in Section 2 and we introduce the tracking procedure used to forecast forward prices. In Section 4 some numerical examples of the solution of the calibration problem and of the performance of the tracking procedure introduced in Section 3 using synthetic and real data are presented.

2. THE STOCHASTIC DYNAMICAL SYSTEM FOR SPIKY ASSET PRICES

Let us introduce a stochastic dynamical system used to model spiky asset prices, see [1,2]. In this model the spiky prices are defined as the product of two stochastic processes: the spike process and the process that describes asset prices in absence of spikes. As said in the Introduction the transitions between spike and regular states are regulated through a two state Markov chain in continuous time.

Let t be a real variable that denotes time, $t \geq 0$, M_t, be the (continuous time) two-state (i.e. regular state, spike state) Markov chain and let

$$P(T,t)=\begin{pmatrix} P_{ss}(T,t) & P_{sr}(T,t) \\ P_{rs}(T,t) & P_{rr}(T,t) \end{pmatrix}, \quad 0<t\leq T, \tag{1}$$

be its 2×2 transition probability matrix. Note that in (1) $P_{ss}(T,t)$ and $P_{rs}(T,t)$ denote the transition probabilities of going from the spike state at time t respectively to the spike or to the regular state at time T, and similarly $P_{sr}(T,t)$

and $P_{rr}(T,t)$ denote the transition probabilities of going from the regular state at time t respectively to the spike or to the regular state at time T.

Note that the Chapman-Kolmogorov equation for the Markov chain M_t, $t \geq 0$, can be written as follows:

$$P(T,t) = P(T,\tau)P(\tau,t),$$
$$\text{for every } t, T, \tau \text{ such that } 0 < t \leq \tau \leq T, \tag{2}$$

together with the condition that $P(T,t)$ is the 2×2 identity matrix when $t = T$.

Let $0 < t \leq T$ and $p_r(t)$, $p_s(t)$, $p_r(T)$, $p_s(T)$ be respectively the probabilities of being in the regular state and in the spike state at time t and at time T. We have:

$$\begin{pmatrix} p_s(T) \\ p_r(T) \end{pmatrix} = P(T,t)\begin{pmatrix} p_s(t) \\ p_r(t) \end{pmatrix}, \quad 0 < t \leq T. \tag{3}$$

In the time-homogeneous case it is possible to write an explicit formula for the matrix $P(T,t)$, $t \leq T$, as a function of the parameters that define the generator of $P(T,t)$, $t \leq T$. These parameters control the duration and the frequency of the spikes, that is they control the expected lifetime of the spike periods and the expected lifetime of the periods between spikes. It is easy to see that in the case of a time-homogeneous Markov chain M_t, $t \geq 0$, the transition probability matrix $P(T,t)$, $0 < t \leq T$, is a function only of the difference $T - t$, and that we have [1]:

$$P(T,t) = \begin{pmatrix} \dfrac{b + ae^{-(T-t)(a+b)}}{a+b} & \dfrac{b - be^{-(T-t)(a+b)}}{a+b} \\ \dfrac{a - ae^{-(T-t)(a+b)}}{a+b} & \dfrac{a + be^{-(T-t)(a+b)}}{a+b} \end{pmatrix}, \quad 0 < t \leq T, \tag{4}$$

where a and b are non negative real parameters. Moreover we assume that a, b are not both zero.

From now on in our model for spiky prices we always assume that the two-state Markov chain M_t, $t \geq 0$, is time-homogeneous and, as a consequence, we assume that its transition probability matrix $P(T,t)$, $0 < t \leq T$, is given by (4).

The process that describes the asset prices in absence of spikes is a diffusion process defined through the following stochastic differential equation:

$$d\hat{S}_t = \mu\hat{S}_t dt + \sigma\hat{S}_t dW_t, \quad t > 0, \tag{5}$$

with the initial condition:

$$\hat{S}_0 = \hat{S}^*, \tag{6}$$

where $\hat{S}_t > 0$ denotes the random variable that describes the asset price in absence of spikes at time t, $t > 0$, μ, σ are constants, $t > 0$, W_t, is the standard Wiener process, $t > 0$, dW_t, $W_0 = 0$, is its stochastic differential and the random variable $\hat{S}^* > 0$ is a given initial condition. For simplicity we assume that the random variable $\hat{S}^*$ is concentrated in a point with probability one and, with abuse of notation, we denote this point with $\hat{S}^*$. Recall that μ is the drift coefficient and $\sigma > 0$ is the volatility coefficient.

Equation (5) defines the asset price dynamics of the celebrated Black-Scholes model. Note that the solution of (5), (6) is a Markov process known as geometric Brownian motion. Note that several other Markov processes different from the one defined in (5), (6) can be used to model asset price dynamics in absence of spikes. For example we can use one of the so called stochastic volatility models that have been introduced recently in mathematical finance to refine the Black-Scholes asset dynamics model.

Let us define the spike process λ_t, $t \geq 0$, that models the amplitude of the spikes in the asset prices. Let ξ_t, $t \geq 0$, be a stochastic process made of independent positive random variables with given probability density functions $\Sigma(t,\xi)$, $\xi > 0$, $t \geq 0$. We assume that if the Markov chain M_t, $t \geq 0$, is in the regular state then the spike process λ_t, $t \geq 0$, is equal to one. If the Markov chain M_t, $t \geq 0$, transits from the regular state to the spike state at time τ, $\tau > 0$, then the spike process λ_t, $t \geq 0$, is equal to a value sampled from the random variable ξ_τ during the entire time period beginning with $t = \tau$ that the Markov chain M_t, $t \geq 0$, remains in the spike state. We assume that M_t, $t \geq 0$, is in the regular state at time $t = 0$ with probability one, so that the spike process λ_t starts with $\lambda_0 = 1$ with probability one. Let us observe that in the case of spikes with constant amplitude $\lambda > 1$, the probability density function

$\Sigma(t,\xi)=\Sigma(\xi),\ \xi>0,\ t\geq 0,$ is the Dirac delta function $\delta(\xi-\lambda)$ with support in $\xi=\lambda,$ that is we have:

$$\Sigma(t,\xi)=\delta(\xi-\lambda),\ \xi>0,\ t\geq 0. \tag{7}$$

Finally we say that the spike process $\lambda_t,\ t\geq 0,$ is in the spike state or in the regular state if the Markov chain $M_t,\ t\geq 0,$ is in the spike state or in the regular state respectively. In our model for spiky prices we assume that the spikes have constant amplitude $\lambda>1,$ that is we assume (7).

Let us define now the process that describes the asset prices with spikes. Let us denote with $S_t>0$ the random variable that models the price (eventually) with spikes of the asset at time $t\geq 0$. We assume that the spike process $\lambda_t,\ t\geq 0,$ and the process $\hat{S}_t,\ t\geq 0,$ that describes asset prices in absence of spikes, are independent. Following [1,2] we define the process $S_t,\ t\geq 0,$ as the product of the spike process $\lambda_t,\ t\geq 0,$ and of the process $\hat{S}_t,\ t\geq 0,$ that is:

$$S_t=\lambda_t\hat{S}_t,\ t\geq 0. \tag{8}$$

Note that the process $S_t,\ t\geq 0,$ for spiky asset prices is in the spike state or in the regular state depending from the fact that the spike process $\lambda_t,\ t\geq 0,$ is in the spike state or in the regular state respectively, or equivalently, depending from the fact that the Markov chain $M_t,\ t\geq 0,$ is in the spike state or in the regular state respectively.

In [1] it has been shown that the process $S_t,\ t>0,$ is able to reproduce spikes in asset prices. This can be seen in Figure 1 where synthetic data sampled from $S_t,\ t>0,$ are shown. These synthetic data are those used in the numerical examples presented in Section 4. Note that depending on the values of the parameters defining the model, the trajectories of $S_t,\ t\geq 0,$ can exhibit a variety of behaviours. In particular, when $M_t,\ t\geq 0,$ is time-homogeneous and $a\neq 0,\ b\neq 0,$ the expected times $\overline{t}_s$ and $\overline{t}_r$ spent by the process $S_t,\ t>0,$ in "spike mode" and in "regular (i.e. inter-spike) mode", respectively, are given by (see [1]):

$$\overline{t_s} = \frac{1}{a}, \ \overline{t_r} = \frac{1}{b}. \tag{9}$$

If $\overline{t_s}$ is small in comparison with the characteristic time of change of the process $\hat{S}_t, \ t > 0,$ then the process $S_t, \ t > 0,$ exhibits spikes. For example, if $\hat{S}_t, \ t > 0,$ is the diffusion process defined in (5) then the previous condition can be stated as follows:

$$\sigma^2 \overline{t_s} = \frac{\sigma^2}{a} \ll 1 \text{ and } \mu \overline{t_s} = \frac{\mu}{a} \ll 1. \tag{10}$$

We interpret $\overline{t_s}$ as the expected lifetime of a spike, and $\overline{t_r}$ as the expected time elapsed between two consecutive spikes. That is the parameters a and b control the duration and the frequency of the spikes through (9). For example, the relations (9) suggest that in order to model short-lived spikes the parameter a must be large, while to model rare spikes the parameter b must be small.

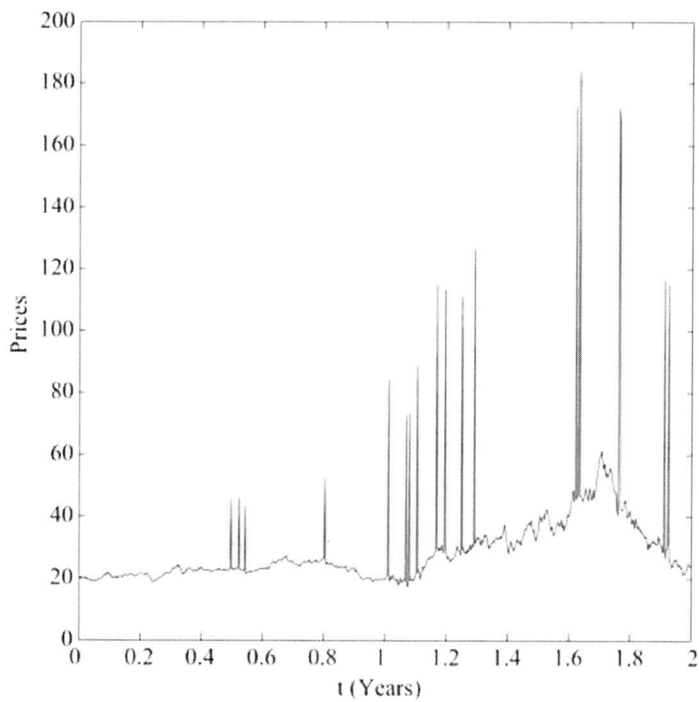

Figure 1. The synthetic data.

Using some heuristic rules to recognize spikes on the data time series considered thank to their relation with the spike duration and frequency a rough estimate of the parameters a and b can be obtained directly from observed spiky prices (market data). For example, the relations (9) suggest that initial estimates a^0 and b^0 of a and b respectively, can be obtained as follows:

$$a^0 = \frac{1}{\overline{t}_s^0}, \quad b^0 = \frac{1}{\overline{t}_r^0}, \tag{11}$$

where $\overline{t}_s^0$ and $\overline{t}_r^0$ are computed (using some heuristic rules) directly from the data available as the observed average lifetime of the spikes and the observed average time between two consecutive spikes, respectively.

Finally it is easy to see that the asymptotic probabilities of the time-homogeneous Markov chain M_t, $t \geq 0$, of being in the spike and in the regular states are respectively:

$$p_s(\infty) = \pi_s = \frac{b}{a+b}, \quad p_r(\infty) = \pi_r = \frac{a}{a+b}. \tag{12}$$

With the previous choices the model for spiky asset prices introduced in [1,2] depends on five parameters, that is: $\mu, \sigma, a, b, \lambda$. The parameters μ and σ are those appearing in (5) and are relative to the asset price model in absence of spikes, the remaining parameters are relative to the spike process (i.e. λ) and to the Markov chain that regulates the transitions between spike and regular states (i.e. a, b).

3. THE FILTERING AND THE CALIBRATION PROBLEMS

The parameters $\mu, \sigma, a, b, \lambda$ are grouped in the following vector:

$$\underline{\Theta} = (\mu, \sigma, a, b, \lambda)^T, \tag{13}$$

where the superscript T denotes the transpose operator. The vector $\underline{\Theta}$ is the unknown of the calibration problem that must be determined from the data available. The vectors $\underline{\Theta}$, describing admissible sets of parameter values, must satisfy some constraints, that is we must have $\underline{\Theta} \in \mathrm{M}$ where:

$$\mathbf{M} = \left\{ \underline{\Theta} = (\mu,\ \sigma,\ a,\ b,\ \lambda)^T \in \mathbb{R}^5 \middle| \sigma \geq 0,\ a \geq 0,\ b \geq 0,\ \lambda \geq 1 \right\}, \tag{14}$$

and $\mathbb{R}^5$ is the five dimensional real Euclidean space.

The constraints contained in $\mathbf{M}$ express some elementary properties of the model introduced in Section 2. Note that, when necessary to take care of special circumstances, more constraints can be added to $\mathbf{M}$.

The data of the calibration problem are the observation times: $0 = t_0 < t_1 < t_2 < \cdots < t_n < +\infty$, and the spiky asset price $\tilde{S}_i$ observed at time t_i, $i = 0, 1, \cdots, n$, where $\tilde{S}_0 = \hat{S}_0 = \hat{S}^*$. The observed price $\tilde{S}_i$ is a positive real number that is understood as being sampled from the random variable S_{t_i}, $i = 0, 1, \cdots, n$. Starting from these data we want to solve the following problems:

Calibration Problem: Find an estimate of the vector $\underline{\Theta} = (\mu, \sigma, a, b, \lambda)^T$.

Filtering Problem (Forecasting Problem): Given the value of the vector $\underline{\Theta} = (\mu,\ \sigma,\ a,\ b,\ \lambda)^T$ estimated forecast the forward asset prices.

Note that the prices $\tilde{S}_j$, $j = 0, 1, \cdots, n$, are called spot asset prices and that with forward asset prices we mean prices "in the future", that is, in our setting, we mean the forward prices at time t_i associated to the spot price $\tilde{S}_i$, $i = n$. These are the prices at time t_i of the asset with delivery time $t > t_i$, $i = n$. Let us point out that in correspondence of a spot price we can forecast several forward prices associated to different delivery periods. The delivery periods of interest are those actually used in real markets, in fact for these delivery periods observed forward prices are available and a comparison between observed forward prices and forecasted forward prices is possible. In Section 4 in the analysis of real data we consider (some of) the delivery periods used in the UK electricity market in the years 2004-2009 and we make this comparison.

Let us mention that the forecasted forward asset prices associated to the spiky price model introduced in Section 2 do not exhibit spikes while, of course, the corresponding (spot) asset prices exhibit spikes (see [1,2] for more details).

The calibration problem consists in estimating the value of the vector $\underline{\Theta}$ that makes "most likely" the observations. The observations available at time $t > 0$ are denoted with:

$$\mathrm{F}_t = \left\{ \tilde{S}_i : t_i \le t \right\}, \quad t > 0. \tag{15}$$

Note that for simplicity we assume that the transitions from regular state to spike state or viceversa happen at the observation times.

Let $p\left(\hat{S},t \middle| \mathrm{F}_t,\underline{\Theta}\right)$, $\hat{S} > 0$, $t > 0$, be the probability density function of the stochastic process $\hat{S}_t$ at time $t > 0$ conditioned to the observations F_t and let $t_i < t \le t_{i+1}$, $\hat{S} > 0$, $p_i\left(\hat{S},t \middle| \underline{\Theta}\right) = p\left(\hat{S},t \middle| \mathrm{F}_{t_i},\underline{\Theta}\right)$, be the probability density function of the stochastic process $\hat{S}_t$ conditioned to the observations made up to time $t = t_i$, when $t_i < t \le t_{i+1}$, $i = 0,1,\cdots,n$, where for convenience we define $t_{n+1} = +\infty$.

The probability density functions $p_i\left(\hat{S},t \middle| \underline{\Theta}\right)$, $\hat{S} \ge 0$, $t_i < t \le t_{i+1}$, $i = 0,1,\cdots,n$, $\underline{\Theta} \in \mathrm{M}$, are the solutions of the following initial value problems for the Fokker-Planck equation associated to the Black-Scholes model (5):

for $i = 0,1,\cdots,n$,

$$\frac{\partial p_i}{\partial t} = \frac{1}{2}\sigma^2 \hat{S}^2 \frac{\partial^2 p_i}{\partial \hat{S}^2} + \mu \hat{S} \frac{\partial p_i}{\partial \hat{S}},$$
$$\hat{S} \ge 0, \quad t_i < t \le t_{i+1}, \quad \underline{\Theta} \in \mathrm{M}, \tag{16}$$

$$p_i\left(\hat{S},t_i \middle| \underline{\Theta}\right) = f_i\left(\hat{S};\underline{\Theta}\right), \quad \hat{S} \ge 0, \quad \underline{\Theta} \in \mathrm{M}, \tag{17}$$

where

$$f_0\left(\hat{S};\underline{\Theta}\right) = \delta\left(\hat{S} - \hat{S}^*\right), \quad \hat{S} > 0, \quad \underline{\Theta} \in \mathrm{M}, \tag{18}$$

$$f_i\left(\hat{S};\underline{\Theta}\right) = p_r(t_i)\delta\left(\hat{S} - \tilde{S}_i\right) + p_s(t_i)\delta\left(\hat{S} - \frac{\tilde{S}_i}{\lambda}\right),$$
$$\hat{S} > 0, \quad i = 1,2,\cdots,n, \quad \underline{\Theta} \in \mathrm{M}, \tag{19}$$

where $p_r(t_i)$ and $p_s(t_i)$ are, respectively, the probabilities defined through the time-homogeneous Markov chain M_t, $t \ge 0$, of being in the regular and in the spike state at time $t = t_i$, $i = 0,1,\cdots,n$ (see (3)).

The probability density functions p_i, $i=0,1,\cdots,n,$ solutions of the initial value problems for the FokkerPlanck equations (16)-(19), can be written as follows:

$$p_i\left(\hat{S},t\middle|\underline{\Theta}\right)=\int_0^{+\infty}\mathrm{d}\hat{S}' p_f\left(\hat{S},t,\hat{S}',t_i\right)f_i\left(\hat{S}';\underline{\Theta}\right),$$
$$\hat{S}\geq 0,\ t_i<t\leq t_{i+1},\ i=0,1,\cdots,n,\ \underline{\Theta}\in \mathrm{M}, \tag{20}$$

where p_f is the fundamental solution of the FokkerPlanck equation (16) associated to the Black-Scholes model (5), that is:

$$p_f\left(\hat{S},t,\hat{S}',t'\right)=\frac{1}{\sigma\hat{S}'\sqrt{2\pi(t-t')}}$$
$$\times e^{-\frac{\left[\log\left(\frac{\hat{S}}{\hat{S}'}\right)-\left(\mu-\frac{1}{2}\sigma^2\right)(t-t')\right]^2}{2\sigma^2(t-t')}},$$
$$\hat{S}\geq 0,\hat{S}'\geq 0,\ t,\ t'\geq 0,\ t-t'>0. \tag{21}$$

In order to measure the likelihood of the vector $\underline{\Theta}$ we introduce the following (log-)likelihood function:

$$F\left(\underline{\Theta}\right)=\sum_{i=0}^{n-1}\log p_i\left(\tilde{S}_{i+1},t_{i+1}\middle|\underline{\Theta}\right),\quad \underline{\Theta}\in\mathrm{M}. \tag{22}$$

It is worthwhile to note that definition (22) contains an important simplification. In fact a more correct definition of the (log-)likelihood function should be:

$$\tilde{F}\left(\underline{\Theta}\right)=\sum_{i=0}^{n-1}\log p_i\left(\tilde{\tilde{S}}_{i+1},t_{i+1}\middle|\underline{\Theta}\right),\quad \underline{\Theta}\in\mathrm{M}, \tag{23}$$

where $\tilde{\tilde{S}}_{i+1}=\tilde{S}_{i+1}$ or $\tilde{\tilde{S}}_{i+1}=\frac{\tilde{S}_{i+1}}{\lambda}$ depending on whether at time t_{i+1} the asset price is in the regular state or in the spike state respectively, $i=0,1,\cdots,n-1.$ However, since when dealing with real financial data the decision about the character of the state (regular or spike state) of the observed prices is dubious, we prefer to adopt the definition (22) for the (log-)likelihood function. In fact the choice made in (22), at the price of introducing some inaccuracy, avoids the necessity of defining a (dubious) criterion to recognize regular and spike states in order to evaluate the (log-)likelihood function. The validity of this choice is supported by the fact that in the numerical experience shown in Section 4 the simplification introduced in (22) using $\tilde{S}_{i+1}$ instead of $\tilde{\tilde{S}}_{i+1}$, $i = 0,1,\cdots,n-1$, is sufficient to obtain satisfactory results.

The solution of the calibration problem is given by the vector $\underline{\Theta}$ that solves the following optimization problem:

$$\max_{\underline{\Theta} \in \mathbf{M}} F(\underline{\Theta}). \tag{24}$$

Problem (24) is called maximum likelihood problem. In fact the vector $\underline{\Theta}^* \in \mathbf{M}$ solution of (24) is the vector that makes "most likely" the observations actually made. Problem (24) is an optimization problem with nonlinear objective function and linear inequality constraints.

Note that (22), (24) is one possible way of formulating the calibration problem considered using the maximum likelihood method. Many other formulations of the calibration problem are possible and legitimate. Moreover the formulation of the calibration problem (22), (24) can be easily extended to handle situations where we consider calibration problems associated to data set different from the one considered here, such as, for example, data set containing asset and option prices or only option prices.

The optimization algorithm used to solve the maximum likelihood problem (22), (24) is based on a variable metric steepest ascent method. The variable metric technique is used to handle the constraints.

Let $|\cdot|$ be the absolute value of $\cdot$, and $\|\cdot\|$ be the Euclidean norm of the vector $\cdot$. Beginning from an initial guess:

$$\underline{\Theta}^0 = \left(\mu^0,\ \sigma^0,\ a^0,\ b^0,\ \lambda^0\right)^T \in \mathbf{M}, \tag{25}$$

we update at every iteration the current approximation of the solution of the optimization problem (24) with a step in the direction of the gradient with respect to $\underline{\Theta}$ computed in a suitable variable metric of the (log-)likelihood function (22).

In particular let us fix a tolerance value $tol > 0$ and a maximum number of iterations $iter > 0$. Given $\underline{\Theta} = \underline{\Theta}^0 \in \mathbf{M}$ the optimization procedure can be summarized in the following steps:

1) Set $k = 0$ and initialize $\underline{\Theta} = \underline{\Theta}^0$;

2) Evaluate $F(\underline{\Theta}^k)$. If $k > 0$ and $\left|F(\underline{\Theta}^k) - F(\underline{\Theta}^{k-1})\right| < tol,$ go to item 7;

3) Evaluate the gradient of the (log-)likelihood function in $\underline{\Theta} = \underline{\Theta}^k$, that is:

$$\nabla F(\underline{\Theta}^k) = \left(\frac{\partial F}{\partial \mu}, \frac{\partial F}{\partial \sigma}, \frac{\partial F}{\partial a}, \frac{\partial F}{\partial b}, \frac{\partial F}{\partial \lambda}\right)^T (\underline{\Theta}^k). \tag{26}$$

If $\left\|\nabla F(\underline{\Theta}^k)\right\| < tol$ go to item 7;

4) Perform the steepest ascent step, evaluating $\underline{\Theta}^{k+1} = \underline{\Theta}^k + \eta_k \nabla F(\underline{\Theta}^k)$, where η_k is a quantity that is used to define the the step made. The quantity η_k can be chosen as a scalar or, more in general, as a matrix of suitable dimensions. The choice of η_k involves the use of the "variable metric". When η_k is chosen to be a scalar we have a "classical" steepest ascent method;

5) If $\left\|\underline{\Theta}^{k+1} - \underline{\Theta}^k\right\| < tol,$ go to item 7;

6) Set $k = k+1$, if $k < iter$ go to item 2;

7) Set $\underline{\Theta}^* = \underline{\Theta}^k$ and stop.

The vector $\underline{\Theta}^*$ obtained in step 7 is the (numerically computed) approximation of the maximizer of the (log-) likelihood function.

Numerical experiments have shown that the (log-) likelihood function $F(\underline{\Theta})$ defined in (22) is a flat function. That is there are many different vectors $\underline{\Theta}$ in M that make likely the data. In the optimization of objective functions with flat regions when local methods, such as the steepest ascent method, are used to solve the corresponding optimization problems (i.e.: (24)), special attention must be paid to the choice of the initial guess of the optimization procedure. That is in actual computations a "good" initial guess $\underline{\Theta}^0 \in \mathrm{M}$ is important to find a "good" solution of the optimization problem (24). More specifically in the problem considered here the use of good initial guesses of the volatility σ and of the drift μ improves substantially the quality of the estimates of all the parameters contained in the vector $\underline{\Theta}$. That is in the solution of problem (24) the parameters μ and σ are the most "sensitive" parameters of the vector $\underline{\Theta}$.

For these reasons in order to improve the robustness of the solution of problem (24) that is obtained with the optimization method described above it is useful to introduce some ad hoc steps that lead to a simple reformulation of the (log-)likelihood function (22), of the calibration problem (24) and of the method used to solve problem (24). First of all an ad hoc preliminary simplified optimization problem is solved to produce a high quality initial guess for the steepest ascent method summarized in steps 1) - 7). We proceed as follows. First we estimate two initial values of μ and σ, let us say $\bar{\mu}$ and $\bar{\sigma}$, directly from the data available, that is we compute the historical volatility and the historical drift of the data available (see, for example, [11]). These estimates are used as initial guesses to maximize the objective function (22) as a two variables function of the parameters μ and σ, with the constraint $\sigma > 0$. Note that in this preliminary step we consider as data used to define the (log-)likelihood function (22) only the "most regular portion of the input data". The most regular portion of the input data is chosen using elementary empirical rules that try to find one or more regular periods in the data considered. In this preliminary optimization problem the remaining components of the vector $\underline{\Theta}$ are initialized as follows $a = b = \bar{a}^0 = \bar{b}^0 = 0$ and $\lambda = \bar{\lambda}^0 = 1$ and these initial values are kept constants in the optimization procedure with respect to μ and σ. Let $(\bar{\mu}^*, \bar{\sigma}^*)$ be the maximizer found at the end of this preliminary step. In order to keep memory in the maximum likelihood problem (24) of the values $\bar{\mu}^*$ and $\bar{\sigma}^*$ found in the preliminary step we add to the objective function (22) the following penalization term:

$$k_1 \left|\mu - \bar{\mu}^*\right|^2 + k_2 \left|\sigma - \bar{\sigma}^*\right|^2, \; k_1 > 0, k_2 > 0, \tag{27}$$

where k_1, k_2 are positive penalization constants. That is we consider the following modified (log-)likelihood function:

$$\begin{aligned} \bar{F}(\underline{\Theta}) &= F(\underline{\Theta}) + k_1 \left|\mu - \bar{\mu}^*\right|^2 + k_2 \left|\sigma - \bar{\sigma}^*\right|^2 \\ &= \sum_{i=0}^{n-1} \log p_i\left(\tilde{S}_{i+1}, t_{i+1} \middle| \underline{\Theta}\right) \\ &\quad + k_1 \left|\mu - \bar{\mu}^*\right|^2 + k_2 \left|\sigma - \bar{\sigma}^*\right|^2, \\ & k_1 > 0, \; k_2 > 0, \; \underline{\Theta} \in \mathrm{M}. \end{aligned} \tag{28}$$

The variable metric steepest ascent method 1) - 7) is now applied to solve problem (24) when the (five variables) (log-)likelihood function $\bar{F}$ defined in

(28) replaces the (log-)likelihood function F defined in (22) starting from the initial guess:

$$\underline{\Theta}^0 = \left(\bar{\mu}^*, \bar{\sigma}^*, a^0, b^0, \lambda^0\right)^T \in \mathrm{M}, \tag{29}$$

where the initial values a^0 and b^0 are estimated directly from the input data through (11) and, finally, the initial value λ^0 is chosen as a rough estimate of the average amplitude of the spikes appearing in the input data. That is problem (24) is solved replacing F with $\bar{F}$ starting from an initial guess that uses the results obtained in the preliminary step of the optimization procedure and tries to exploit the data using the "physical" meaning of the model parameters.

These ad hoc steps used to reformulate problem (24), when applied to the analysis of synthetic data, lead to a substantial improvement in the accuracy of the estimates of the model parameters obtained when compared to the estimates obtained solving directly (22), (24) and to a great saving in the computational cost of solving the calibration problem. This reformulation of problem (24) is used to analyze the data considered in Section 4.

As a byproduct of the solution of the calibration problem we obtain a technique to forecast forward asset prices. Let us consider a filtering problem. We assume that the vector $\underline{\Theta}$ solution of the calibration problem associated to the data $\mathrm{F}_t = \{\tilde{S}_i : t_i \le t\},\ t = t_n > 0$ is known. From the knowledge of $\underline{\Theta} = (\mu, \sigma, a, b, \lambda)^T$ at time $t = t_n$ we can forecast the forward asset prices with delivery period $\Delta t > 0$ deep in the future and delivery time $\tau_n = t_n + \Delta t$ as follows:

$$\begin{aligned} \mathrm{E}\left(S_{\tau_i} \middle| \mathrm{F}_{t_n}, \underline{\Theta}\right) &= \mathrm{E}\left(\lambda_{\tau_i}\right)\mathrm{E}\left(\hat{S}_{\tau_i}\right) \\ &= \left(1 \cdot p_r(\tau_i) + \lambda \cdot p_s(\tau_i)\right)\mathrm{E}\left(\hat{S}_{t_i}\right)e^{\mu(\tau_i - t_i)}, \\ \tau_i &= t_i + \Delta t, \quad i = n, \end{aligned} \tag{30}$$

where $\mathrm{E}(\cdot)$ denotes the expected value of $\cdot$.

In the numerical experiments presented in Section 4 we use the following approximation:

$$\mathrm{E}\left(\hat{S}_{t_i}\right) \approx \frac{1}{10}\sum_{k=0}^{9}\tilde{S}_{i-k}, \quad i = n. \tag{31}$$

Note that using formula (31) we have implicitly assumed $n>9$. In fact in the data time series considered in Section 4 the average in time of the (spiky) observations appearing in (31) gives a better approximation of the "spatial" average $\mathrm{E}(\hat{S}_{t_i})$ of the price without spikes than the individual observation $\tilde{S}_i$ made at time $t=t_i$ of the spiky price. However the average in time of the observations approximates the "spatial" average only if short time periods are used. This is the reason why we limit the mean contained in (31) to the data corresponding to ten consecutive observation times, that is corresponding to a period of ten days when we process daily data as done in Section 4. Note that in (31) an average of spiky prices is used to approximate the expected value of non spiky prices.

4. NUMERICAL RESULTS

The first numerical experiment presented consists in solving the calibration problem discussed in Section 3 using synthetic data. This experiment does the analysis of a time series of daily data of the spiky asset price during a period of two years. The time series studied is made of 730 daily data of the spiky asset prices, that is the prices $\tilde{S}_i$ at time t_i, $i=0,1,\cdots,729$. These data have been obtained computing one trajectory of the stochastic dynamical system used to model spiky asset prices defined in Section 2 looking at the computed trajectory at time $t=t_i$, $i=0,1,\cdots,729$, with $t_i-t_{i-1}=1/365$, $i=1,2,\cdots,729$, where we have chosen $t_0=0$ and $\tilde{S}_0=\hat{S}_0=\hat{S}^*=20$. We choose the vector $\underline{\Theta}$ that specifies the model used to generate the data equal to $\underline{\Theta}=(\mu,\ \sigma,\ a,\ b,\ \lambda)^T=\underline{\Theta}_1=(0.03,\ 0.3,\ 70,\ 1,\ 2)^T$ in the first year of data (i.e. when $t=t_i$, $i=0,\ 1,\ \cdots,\ 364$), and equal to $\underline{\Theta}=(\mu,\ \sigma,\ a,\ b,\ \lambda)^T=\underline{\Theta}_2=(0.1,\ 0.8,\ 150,\ 5,\ 4)^T$ in the second year of data (i.e. when $t=t_i$, $i=365,\ 366,\ \cdots,\ 729$). The data are generated using as initial value of the second year the last datum of the first year. The synthetic data generated in this way are shown in Figure 1. These data are spiky data and the fact that the first year of data is generated using a different choice of $\underline{\Theta}$ than the choice made in the second year of data can be seen simply looking at Figure 1.

We solve problem (24) with the ad hoc procedure described in Section 3 using the data associated to a time window made of 365 consecutive observation times, that is 365 days (one year), and we move this window across the two years of data discarding the datum corresponding to the first observation time of the window and inserting the datum corresponding to the next observation time after the window. Note that numerical experiments suggest that it is necessary to take a large window of observations to obtain a good estimate the parameters a, b and λ. The calibration problem is solved for each choice of the data time window applying the procedure described in Section 3, that is it is solved 365 times. The 365 vectors $\underline{\Theta}$ constructed in this way are compared to the two choices of the model parameters $\underline{\Theta}$ used to generate the data. Moreover the time when $\underline{\Theta}$ changes from being $\underline{\Theta}_1$ to being $\underline{\Theta}_2$ is reconstructed.

To represent the 365 vectors obtained solving the calibration problem in correspondence to the 365 data time windows considered, we associate to each reconstructed vector $\underline{\Theta} \in R^5$ a point on a straight line (see Figure 2). Let us explain how this correspondence is established. We first represent the vectors $\underline{\Theta}_1$ and $\underline{\Theta}_2$ that generate the data as two points on the straight line mentioned above having a distance proportional to $\|\underline{\Theta}_1 - \underline{\Theta}_2\|^*$ measured in $\|\underline{\Theta}_1\|^*$ units, where

$\|\underline{\Theta}\|^* = \frac{1}{5}\sum_{i=1}^{5}|\underline{\Theta}(i)|$ and

$\underline{\Theta} = (\Theta(1), \Theta(2), \cdots, \Theta(5))^T \in R^5$. We choose the origin of the straight line to be the mid point of the segment that joins $\underline{\Theta}_1$ and $\underline{\Theta}_2$. In Figure 2 the diamond represents the vector $\underline{\Theta}_1$ and the square represents the vector $\underline{\Theta}_2$. The unit length is $\|\underline{\Theta}_1\|^*$. The vectors (points) solution of the 365 calibration problems are represented as (green or red) stars. A point $P = \underline{\Theta}$ is plotted around $\underline{\Theta}_1$ when the quantity $\frac{\|\underline{\Theta} - \underline{\Theta}_1\|^*}{\|\underline{\Theta}_1\|^*}$ is smaller than the quantity

$\frac{\|\underline{\Theta} - \underline{\Theta}_2\|^*}{\|\underline{\Theta}_2\|^*}$, otherwise the point P is plotted around $\underline{\Theta}_2$.

The distance of the point P from $\underline{\Theta}_1$ when P is plotted around $\underline{\Theta}_1$ (or from $\underline{\Theta}_2$ when P is plotted around $\underline{\Theta}_2$) is $\|\underline{\Theta}_1 - \underline{\Theta}\|^*$ measured in $\|\underline{\Theta}_1\|^*$ units (or $\|\underline{\Theta}_2 - \underline{\Theta}\|^*$ measured in $\|\underline{\Theta}_1\|^*$ units). The point $P = \underline{\Theta}$ plotted around $\underline{\Theta}_1$ is plotted to the right or to the left of $\underline{\Theta}_1$ according to the sign of the second component of $\underline{\Theta} - \underline{\Theta}_1$ (negative second component of $\underline{\Theta} - \underline{\Theta}_1$ is plotted to the left of $\underline{\Theta}_1$). Remind that the second component of $\underline{\Theta}$ is the volatility coefficient σ. A similar statement holds for the points plotted around $\underline{\Theta}_2$. The results obtained in this experiment are shown in Figure 2. In this figure the green stars represent the solutions of the calibration problems associated to the first 183 data time windows, that is the first "half" of data time windows, while the red stars represent those associated to the second "half" of the data time windows (that is the second

182 time windows). Figure 2 shows that the points (vectors) obtained solving the calibration problems are concentrated on two disjoint segments one to the left and one to the right of the origin and that they form two disjoint clusters around $\underline{\Theta}_1$ and $\underline{\Theta}_2$. That is, the solution of the 365 calibration problems corresponding to the 365 time windows described previously shows that two sets of parameters seems to generate the data studied. This is really the case. Moreover, as expected, the points of the cluster around $\underline{\Theta}_2$ are in majority red stars, that is they are in majority the points obtained by the analysis of data time windows containing a majority of observations made in the second year (the second "half" of the data time windows), and a similar statement holds for the points of the cluster around $\underline{\Theta}_1$ (in majority green stars).

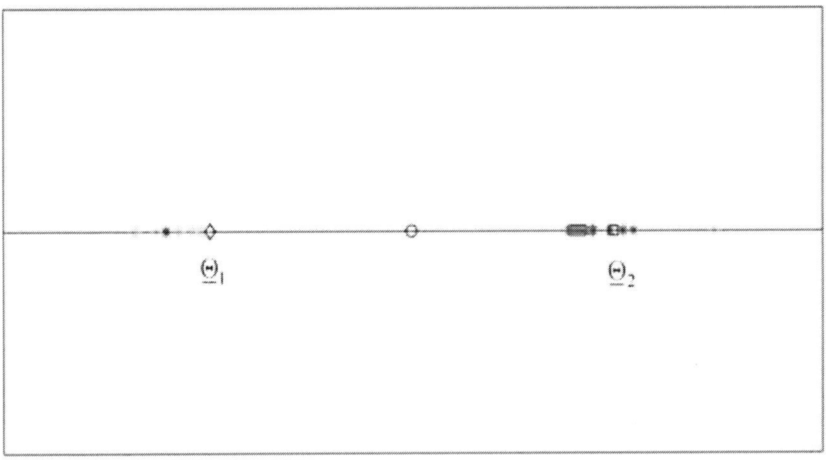

Figure 2. Reconstruction of the parameter vectors $\underline{\Theta}_1$ and $\underline{\Theta}_2$.

The second numerical experiment is performed using real data. The real data studied are electric power price data taken from the UK electricity market. These data are "spiky" asset prices. The data time series studied is made of the observation times $0 = t_0 < t_1 < t_2 < \cdots < t_n = 1395 < +\infty$, (days), and of the spiky asset price $\tilde{S}_i$ observed at time t_i, where $\tilde{S}_i$ is the daily electric power spot price (GBP/MWh), named Day-Ahead price, $i = 0, 1, \cdots, n$. Excluding week-end days and holidays this data time series corresponds to more than 5 years of daily observations going from January 5, 2004 to July 10, 2009. Remind that GBP means Great Britain Pound and that MWh means Mega-Watt/hour.

Moreover the data time series studied in correspondence of each spot price contains a series of forward prices associated to it for a variety of delivery periods. These prices include: forward price 1 month deep in the future (Month-Ahead price), forward price 3 months deep in the future (Quarter-Ahead price), forward price 4 months deep in the future (Season-Ahead price), forward price 1 year deep in the future (One Year-Ahead price). These forward prices are observed each day t_i and the forward prices observed at time t_i are associated to the spot price $\tilde{S}_i$, $i=0,1,\cdots,1395$. The spot and the forward prices contained in the data time series mentioned above are shown in Figure 3.

The observed electric power prices generate data time series with a complicated structure. The stochastic dynamical system studied in this paper does not pretend to fully describe the properties of the electric power prices. Indeed it is able to model only one property of these prices: the presence of spikes. In addition the electricity prices have many other properties, for example, they are mean reverting and have well defined periodicity, that is they have diurnal, weekly and seasonal cycles. A lot of specific models incorporating (some of) these features are discussed in the literature, see for example [10].

Let us begin performing the analysis of the data using the model introduced in Section 2. The first question to answer is: the model for spiky prices presented in Section 2 is an adequate model to analyze the time series of the spot prices? We answer to this question analyzing the relationship between the data and the model parameters established through the solution of the calibration problem. Let us begin showing that the relation between the data and the model parameters established through the calibration problem is a stable relationship. We proceed as follows. We have more than 5 years of daily observations. We apply the calibration procedure to the data corresponding to a window of 257 consecutive observation times. Note that 257 is approximately the number of working days contained in a year and remind that we have data only in the working days. We move this window through the data time series discarding the datum corresponding to the first observation time of the window and adding the datum corresponding to the next observation time after the window. In this way we have 1396 − 257 = 1139 data windows and for each one of these data windows we solve the corresponding calibration problem. The calibration problems are solved using the optimization procedure (including the ad hoc steps) described in Section 3. The reconstructions of the parameters obtained moving the window along the data are shown in Figures 4 and 5. In Figures 4 and 5 the abscissa represents the data windows used to reconstruct the model parameters numbered in ascending order according to the order in time of the first day of the window. Figures 4 and 5 show that the model parameters, with the exception of λ, are approximately constant functions of the data window. The parameter λ reconstructed shown in Figure 5 is a piecewise constant function. These findings support the idea that the model and the formulation of the calibration problem presented respectively in Sections 2 and 3 are adequate to interpret the data. In fact they establish a stable relationship between the data and the model parameters as shown in Figures 4 and 5.

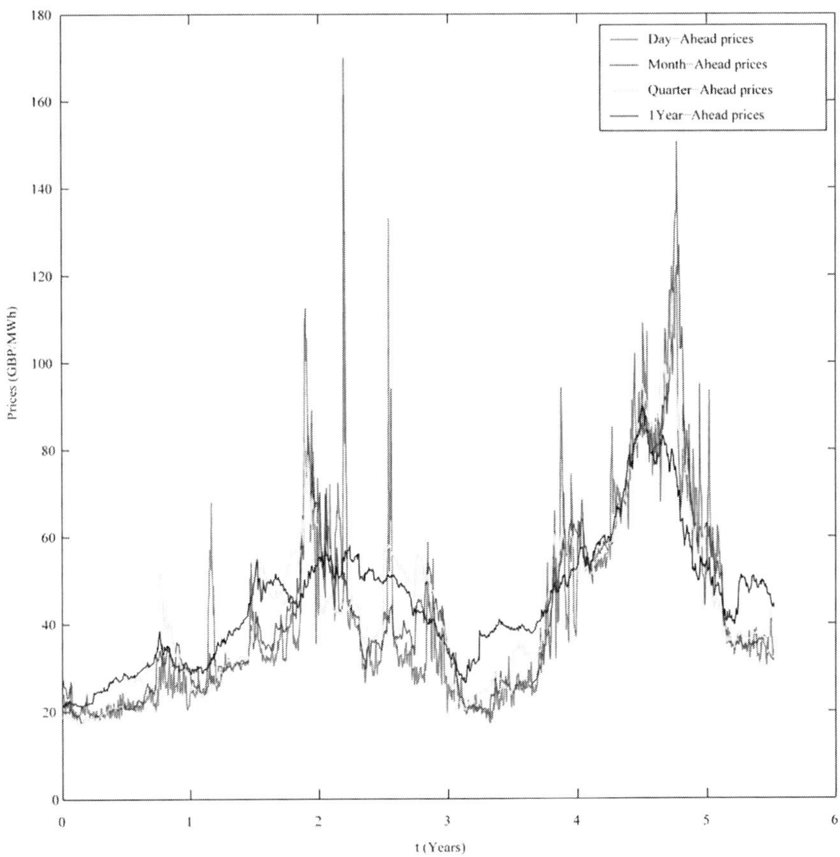

Figure 3. The UK electric power price data.

In the analysis of the real data time series the second question to answer is: the solution of the calibration problem and the tracking procedure introduced in Section 3 can be used to forecast forward prices? To answer this question we compare the observed and the forecasted forward prices. We apply the calibration procedure to a data window made of the first three years of consecutive observations of the spot price taken from the data time series shown in Figure 3 and we use the solution of the calibration problem found and formulae (30), (31) to calculate the forecasted forward prices associated to the spot prices of the data time series not included in the data window mentioned above used in the calibration problem. In Figures 6-8 the forward electric power prices forecasted are shown and compared to the observed forward electric power prices. In Figures 6-8 the abscissa is the day of the spot price associated to the forward prices computed. The abscissa of Figures 6-8 is coherent with the abscissa of Figure 3. Table 1 summarizes quantitatively the results shown in Figures 6-8 and gives the average relative error $e_{\text{forward prices}}$ committed using the forecasted forward prices, that is the average relative error committed approximating the observed forward prices with the forecasted forward prices. Table 1 and Figures 6-8 show the high quality of the forecasted forward prices answering the second question posed about the analysis of the data time series in the affirmative.

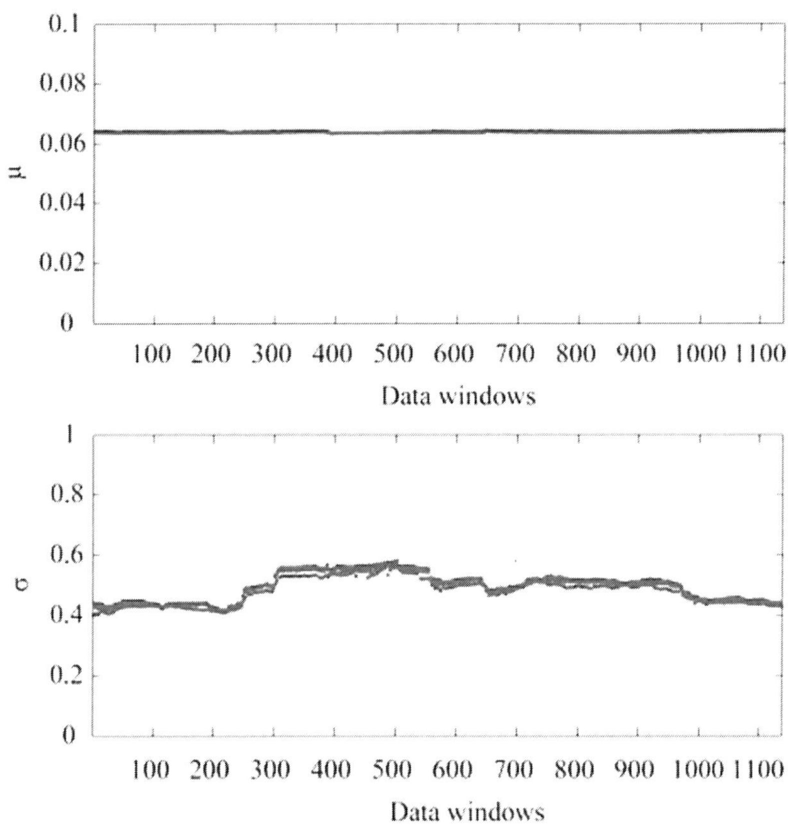

Figure 4. Reconstruction of the parameters of the BlackScholes model: μ, σ.

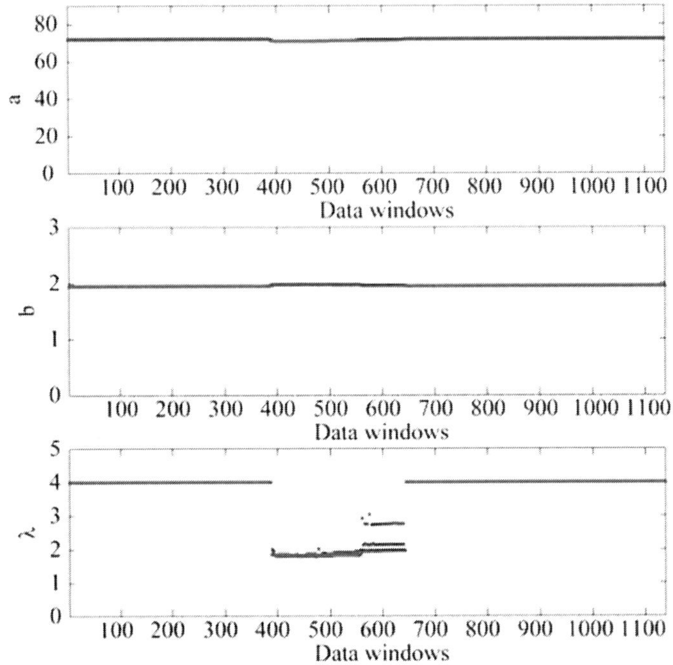

Figure 5. Reconstruction of the parameters used to model the spikes: a, b, λ.

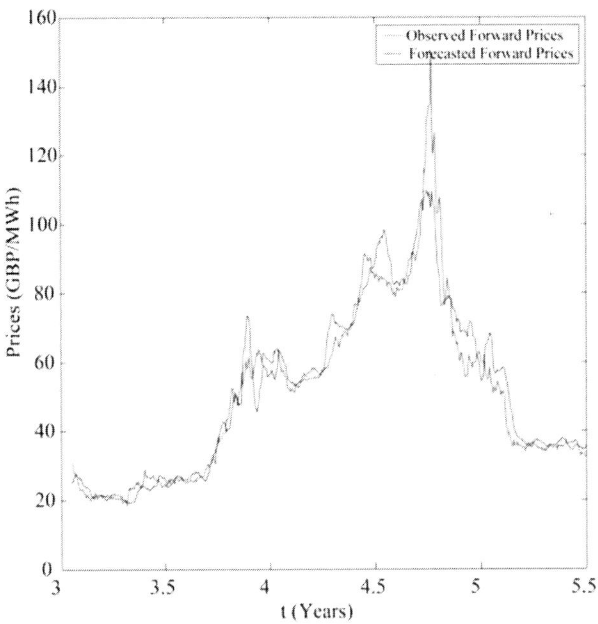

Figure 6. Month-ahead prices.

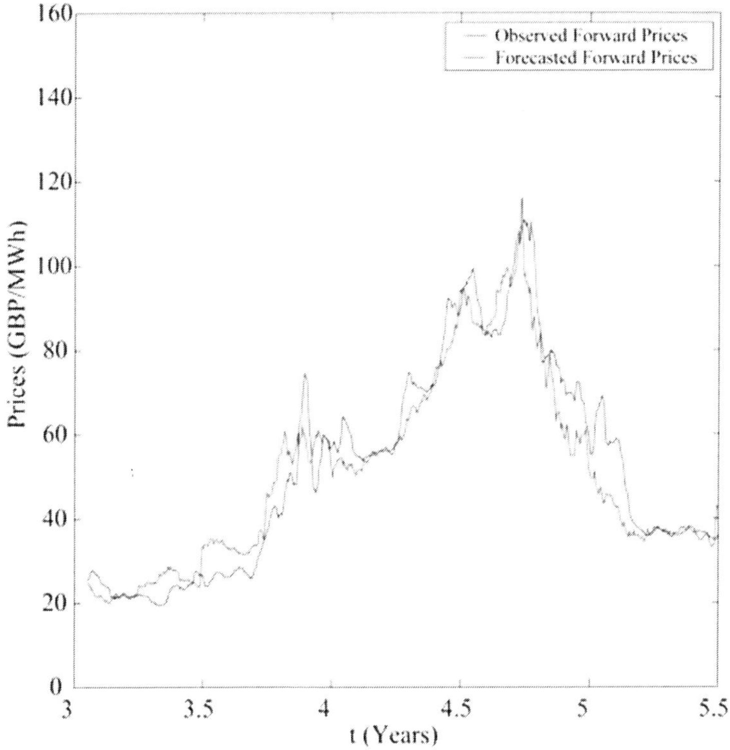

Figure 7. Quarter-ahead prices.

We can conclude that the data analysis presented shows that the model introduced to describe spiky prices, the formulation of the calibration problem and its numerical solution presented in this paper have the potential of being tools of practical value in the analysis of data time series of spiky prices.

Table 1. Average relative errors of the forecasted forward electric power prices when compared to the observed forward prices.

Number of days in the future	$e_{\text{forward prices}}$
30 (Month-Ahead prices)	0.0796
90 (Quarter-Ahead prices)	0.1160
365 (One Year-Ahead prices)	0.2183

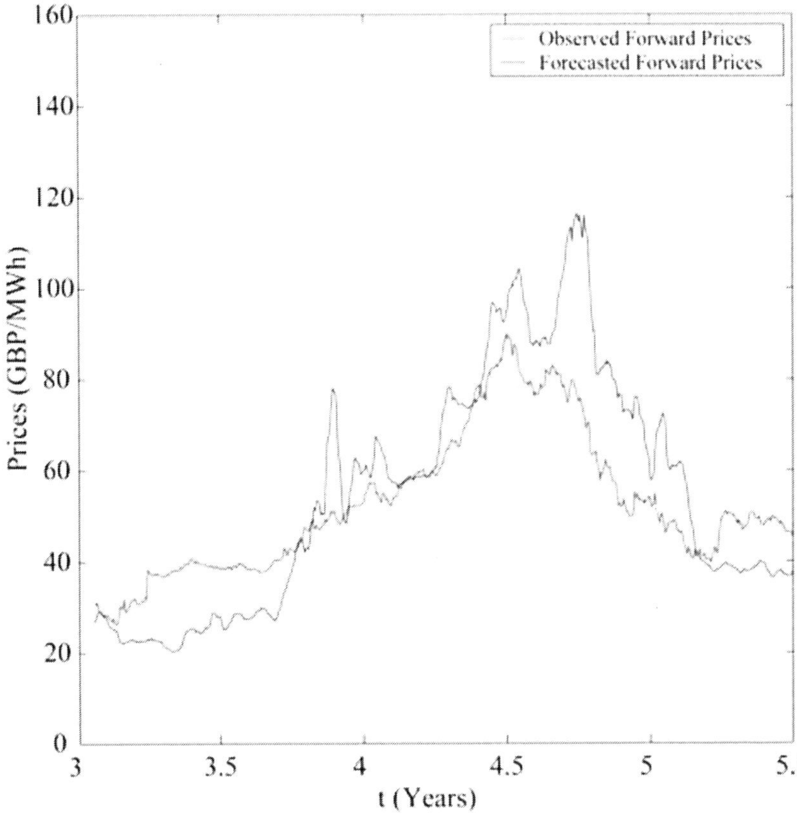

Figure 8. One year-ahead prices.

REFERENCES

1. V. A. Kholodnyi, "Valuation and Hedging of European Contingent Claims on Power with Spikes: A Non-Markovian Approach," Journal of Engineering Mathematics, Vol. 49, No. 3, 2004, pp. 233-252.

2. V. A. Kholodnyi, "The Non-Markovian Approach to the Valuation and Hedging of European Contingent Claims on Power with Scaling Spikes," Nonlinear Analysis: Hybrid Systems, Vol. 2, No. 2, 2008, pp. 285-304.

3. F. Mariani, G. Pacelli and F. Zirilli, "Maximum Likelihood Estimation of the Heston Stochastic Volatility Model Using Asset and Option Prices: An Application of Nonlinear Filtering Theory," Optimization Letters, Vol. 2, No. 2, 2008, pp. 177-222.

4. L. Fatone, F. Mariani, M. C. Recchioni and F. Zirilli, "Maximum Likelihood Estimation of the Parameters of a System of Stochastic Differential Equations that Models the Returns of the Index of Some Classes of Hedge Funds," Journal of Inverse and Ill Posed Problems, Vol. 15, No. 5, 2007, pp. 493-526.

5. L. Fatone, F. Mariani, M. C. Recchioni and F. Zirilli, "The Calibration of the Heston Stochastic Volatility Model Using Filtering and Maximum Likelihood Methods," G. S. Ladde, N. G. Medhin, C. Peng and M. Sambandham, Eds., Proceedings of Dynamic Systems and Applications, Dynamic Publishers, Atlanta, Vol. 5, 2008, pp. 170-181.

6. L. Fatone, F. Mariani, M. C. Recchioni and F. Zirilli, "Calibration of a Multiscale Stochastic Volatility Model Using European Option Prices," Mathematical Methods in Ecomics and Finance, Vol. 3, No. 1, 2008, pp. 49-61.

7. L. Fatone, F. Mariani, M. C. Recchioni and F. Zirilli, "An Explicitly Solvable Multi-Scale Stochastic Volatility Model: Option Pricing and Calibration," Journal of Futures Markets, Vol. 29, No. 9, 2009, pp. 862-893.

8. L. Fatone, F. Mariani, M. C. Recchioni and F. Zirilli, "The Analysis of Real Data Using a Multiscale Stochastic Volatility Model," European Financial Management, 2012, in press.

9. P. Capelli, F. Mariani, M. C. Recchioni, F. Spinelli and F. Zirilli, "Determining a Stable Relationship between Hedge Fund Index HFRI-Equity and S&P 500 Behaviour, Using Filtering and Maximum Likelihood," Inverse Problems in Science and Engineering, Vol. 18, No. 1, 2010, pp. 93-109.

10. C. R. Knittel and M. R. Roberts, "An Empirical Examination of Restructured Electricity Prices," Energy Economics, Vol. 27, No. 5, 2005, pp. 791-817.

11. J. H. Hull, "Options, Futures, and Other Derivatives," 7th Edition, Pearson Prentice Hall, London, 2008.

NOTES

*The research reported in this paper is partially supported by MUR— Ministero Università e Ricerca (Roma, Italy), 40%, 2007, under grant: "The impact of population ageing on financial markets, intermediaries and financial stability". The support and sponsorship of MUR are gratefully acknowledged.

#The numerical experience reported in this paper has been obtained using the computing grid of ENEA (Roma, Italy). The support and sponsorship of ENEA are gratefully acknowledged.

Chapter 10

On the Stochastic Dominance of Portfolio Insurance Strategies

Hela Maalej[1], Jean-Luc Prigent[2]

[1]ThEMA, University of Cergy-Pontoise, Boulevard du Port, France

[2]ThEMA and LabeX MME-DII, University of Cergy-Pontoise, Boulevard du Port, France

ABSTRACT

This paper compares the performance of the two main portfolio insurance strategies, namely the Option-Based Portfolio Insurance (OBPI) and the Constant Proportion Portfolio Insurance (CPPI). For this purpose, we use the stochastic dominance approach. We provide several explicit sufficient conditions to get stochastic dominance results. When taking account of specific constraints, we use the consistent statistical test proposed by Barret and Donald [1] . It is similar to the Kolmogrov- Smirnov test with a complete set of restrictions related to the various forms of stochastic dominance. We find that the CPPI method can perform better than the OBPI one at the third order stochastic dominance.

KEYWORDS

Stochastic Dominance, Portfolio Insurance, CPPI, OBPI, Barret and Donald Test

1. INTRODUCTION

The goal of portfolio insurance is to provide a guarantee against portfolio downside risk (usually 100% of the initial invested amount) while allowing to benefit from significant gains for bullish markets. The two standard portfolio insurance methods are the Option Based Portfolio Insurance (OBPI), introduced by Leland and Rubinstein [2] and the Constant Proportion Portfolio Insurance (CPPI) considered by Perold [3] . Basically, the OBPI portfolio is a combination of a risky asset S (usually a financial index such as the S&P) and a put written on it. Whatever the value of S at the terminal date T, the portfolio value will be always higher than the strike of the put. Therefore this strike is chosen in order to provide the desired guaranteed level. The standard CPPI method consists in a simplified strategy to allocate assets dynamically over time. A floor is initially determined such as it is equal to the lowest acceptable value of the portfolio. Then, the amount allocated to the risky asset (called the exposure) is defined as follows: the cushion, which is equal to the excess of the portfolio value over the floor is multiplied by a predetermined multiple. The remaining funds are usually invested in the reserve asset, usually T-bills. As the cushion approaches zero, exposure approaches zero too. In continuous time, this keeps portfolio value from falling below the floor.

To compare these two main portfolio strategies, we search for stochastic dominance (SD) properties since SD takes account of the entire return distribution. The major argument for stochastic dominance is that it does not require any specific knowledge about the preferences of investors. Indeed, the first stochastic dominance order is related to investors with an increasing utility function. Stochastic dominance of order two focuses on investors having an increasing and concave utility, meaning that they are risk-averse1. However, at a given order (for example 1 or 2), the stochastic dominance criterion cannot always allow to rank all portfolios. There exist cases where no stochastic dominance is observable. But there exists a stochastic dominance criteria at each order and, the higher the order, the less stringent the criterion. Thus it is reasonable to expect that there exists an order for which a portfolio strategy dominates another one (or vice versa). De Giorgi [6] shows that, in a market without friction, the market portfolio can be efficient according to the criterion of the second order stochastic dominance. Therefore the test of stochastic dominance is consistent with the theory of portfolio choice. To compare with alternative approaches such as those based on performances measures, note that Darsinos and Satchell [7] show that n-order stochastic dominance implies Kappa (n – 1) dominance. It means for example that the second order stochastic dominance implies the Omega dominance while the third order SD implies dominance according to the Sortino measure.

For the portfolio insurance strategies, Bertrand and Prigent [8] proved that the stochastic dominance at the first order is a too strong condition, meaning that neither the CPPI nor the OBPI dominates the other strategy for this criterion2. However, as proved theoretically by Zagst and Kraus [10] , stochastic dominance of portfolio insurance strategies can be obtained mainly from the third order. Our main purpose is to extend previous results when taking account of quite general share values and/or of specific constraints such as capped strategies introduced to limit financial risk exposures.

The paper is organized as follows. In Section 2, we briefly introduce the basic properties of the CPPI and the OBPI strategies. In Section 3, we examine the stochastic dominance (SD) framework to compare portfolio insurance strategies. First, we provide several sufficient conditions to get stochastic dominance properties for the standard portfolio insurance methods. Second, to extend previous results, we introduce specific statistical tests and simulation methods for computing p-values when examining SDj with j larger than one. We use the test considered by Barret and Donald [1] , based on the multiplier central limit theory discussed in Van der Vaart and Wellner [11] .

2. BASIC PROPERTIES OF THE CPPI AND THE OBPI STRATEGY

2.1. The Financial Market

We consider two basic assets that are traded in continuous time during the investment period $[0,T]$. The “risk-free” asset (a money market account for example) is denoted by B. Denote by r the constant continuous interest rate $r > 0$. We get:

$$B_t = B_0 \cdot e^{rT}, \tag{1}$$

with initial value $B_0 > 0$. The risky asset (for example a financial market index) is denoted by S. It is assumed to be a geometric Brownian motion given by:

$$dS_t = S_t\left(\mu dt + \sigma dW_t\right), \tag{2}$$

with non negative initial value $S_0 > 0$ and where $W = (W_t)_{(0\le t\le T)}$ is a standard Brownian motion. There exists a constant drift term parameterized by $\mu > r$ and a volatility denoted by σ.

To price options, we use the Black and Scholes formula while taking account of the spread between the empirical and the implied volatility3.

2.2. Constant Proportion Portfolio Insurance (CPPI)

The standard CPPI method consists in a simplified strategy to allocate assets dynamically over time so that its value V_t^{CPPI} never falls below the floor F_t. This latter one is equal to the lowest acceptable value of the portfolio and is defined as a percentage p ($1 \geq p \geq 0$) of the initial investment V_0^{CPPI}, i.e.:

$$F_t = pe^{rt} V_0^{CPPI}. \tag{3}$$

The excess of the portfolio value over the floor is called the cushion C_t. It is equal to:

$$C_t = \max\left\{V_t^{CPPI} - F_t, 0\right\}. \tag{4}$$

Then, the amount allocated on the risky asset (called the exposure E_t) is equal to the cushion multiplied by a constant multiple m. Therefore, the exposure $(E_t)_{(0\leq t\leq T)}$ satisfies:

$$E_t = m \cdot C_t = m \cdot \max\left\{V_t^{CPPI} - F_t, 0\right\}. \tag{5}$$

The interesting case is when $m > 1$, that is when the payoff function of the portfolio value at maturity V_T^{CPPI} is a convex function with respect to the risky asset S_T.

Then the cushion value C_t must satisfy:

$$\mathrm{d}C_t = \mathrm{d}(V_t - F_t) = (V_t - E_t)\frac{\mathrm{d}B_t}{B_t} + E_t\frac{\mathrm{d}S_t}{S_t} - \mathrm{d}F_t.$$

By applying Itô's lemma, we obtain:

$$C_t = C_0 \cdot \exp\left[\left(r + m(\mu - r)\right)t - \frac{1}{2}m^2\sigma^2 t\right] \cdot \exp\left[m\sigma W_t\right]. \tag{6}$$

By using the relation:

$$S_t^m = S_0^m \cdot \exp\left[m\sigma W_t + m\left(\mu - \frac{1}{2}\sigma^2\right)t\right],$$

we deduce:

$$\exp[m\sigma W_t] = \frac{S_t^m}{S_0^m}\exp\left[-m\left(\mu t - \frac{1}{2}m^2\sigma^2 t\right)\right].$$

Substituting this expression for $\exp[m\sigma W_t]$ into the expression for C_t leads to:

$$C_t(m, S_t) = C_0 \cdot \left(\frac{S_t}{S_0}\right)^m \exp\left[rt(1-m) + \frac{1}{2}m\sigma^2 t - \frac{1}{2}m^2\sigma^2 t\right] = \alpha_t S_t^m, \tag{7}$$

with

$$\alpha_t = C_0 \cdot \left(\frac{1}{S_0}\right)^m \exp\left[rt(1-m) + \frac{1}{2}m\sigma^2 t - \frac{1}{2}m^2\sigma^2 t\right].$$

We deduce that the value of the CPPI portfolio V_t^{CPPI} at any time t is given by:

$$V_t^{CPPI}(m, S_t) = F_t e^{rt} + \alpha_t S_t^m. \tag{8}$$

Note that, for the CPPI method, the two key management parameters are the initial floor value $F_0 = pV_0 e^{-rT}$ and the multiple m.

Remark 2.1. (Capped CPPI) The manager may want to increase his profits, from usual performances of asset S while potentially discarding very high values of S. In that case, the exposure is determined by:

$$E_t = \inf\left\{mC_t, \lambda V_t^{CPPI}\right\}, \tag{9}$$

where λ denotes the gearing coefficient. Its usual value is equal to 0.9.

2.3. Option-Based Portfolio Insurance (OBPI)

In what follows, we describe the option-based portfolio insurance strategy. It provides a guarantee equal to $p \cdot V_0^{OBPI}$ whatever the market fluctuations. Indeed, for a given share q, we have:

$$V_T^{OBPI} = q\left[K + (S_T - K)^+\right], \tag{10}$$

which implies that $V_T^{OBPI} \geq p \cdot V_0^{OBPI}$ if $qK = p \cdot V_0^{OBPI}$.

This relation shows that the insured amount at maturity is the exercise price multiplied by the number of shares, i.e. qK. The value V_T^{OBPI} of this portfolio at any time t in the period $[0,T]$ is equal to:

$$V_t^{OBPI} = q\left[Ke^{(-r(T-t))} + C(t,K)\right],$$

where $C(t,K)$ denotes the Black-Scholes value of the European call option with strike K, calculated under the risk neutral probability Q.

The portfolio value V_t^{OBPI}, for all dates t before T, is always above the deterministic level $qKe^{-r(T-t)}$. In order to guarantee the minimum terminal portfolio value $p \cdot V_0^{OBPI}$, the strike K of the European Call option must satisfy the following relation:

$$p \cdot V_0 = qK,$$

which implies that:

$$\frac{C(0,K)}{K} = \frac{1-pe^{-rT}}{p}. \tag{11}$$

Therefore, the strike K is an increasing function $K(p)$ of the percentage p, since in Equation (Equation (11)) both functions are decreasing respectively with respect to K and p. Then, the number of shares q is given by:

$$q = \frac{V_0}{Ke^{(-rT)} + C(0,K)}. \tag{12}$$

Thus, for any investment value V_0, the number of shares q is a decreasing function of the percentage p.

In what follows, we price the option using the implicit volatility σ_i.

We denote its price by $Call(S_0, K, r, \sigma^i)$.

Remark 2.2. (Capped OBPI) If the manager wants to increase his profit while potentially discarding very high value of S, the options are capped at a level L, as follows. Consider a parameter L higher than the strike K.

The terminal value of the capped OBPI with strike K and parameter L is defined by:

$$V_{cap,T}^{OBPI} = qMin\left[K + (S_T - K)^+, L\right],$$
$$= q\left[K + (S_T - K)^+ - (S_T - L)^+\right]. \tag{13}$$

3. STOCHASTIC DOMINANCE OF PORTFOLIO INSURANCE STRATEGIES

3.1. Stochastic Dominance: Theoretical Results

In what follows, we provide several sufficient conditions to get stochastic dominance results as in Zagst and Kraus [10] but without assuming as them that q is equal to 1 (see previous Relation 12).

3.1.1. The Second Order Stochastic Dominance

Mosler [12] has stated a theorem for determining the second order stochastic dominance (denoted by $\succ_2$) between random variables based on the condition of intersection between the cumulative distribution functions.

Theorem 3.1. (Mosler [12]). Let V^* and V be two random variables with finite expectations. Denote for all $x \in [a,b]$, $H(x) = F_{V^*}(x) - F_V(x)$ the difference of their respective cumulative distributions functions. Then, we get:

$$H \in \mathrm{S}_1,\ E[V] \leq E[V^*] \Rightarrow V^* \succ_2 V,$$

where S_k denotes the set of all real functions H, with k changes of sign:

$$\mathrm{S}_k = \Big\{H : \mathrm{R} \to \mathrm{R} : \exists s_1, \cdots, s_k \in \mathrm{R},\ s_0 := -\infty, s_{k+1} := +\infty$$
$$\text{with } (-1)^j \cdot H(s) \geq 0, \forall s \in (s_j, s_{j+1}),\ j = 0, \cdots, k,\ H \neq 0\Big\}.$$

We deduce that, if $H(x) \in \mathrm{S}_k$, then the two functions $F_{V^*}(x)$ and $F_V(x)$ intersect k times.

For example, we have:

$$\mathrm{S}_1 = \left\{H : \mathrm{R} \to \mathrm{R} : \exists\, s_1 \in \mathrm{R} \text{ with } H(s) \begin{cases} \geq 0, s \in (-\infty, s_1) \\ \leq 0, s \in (s_1, +\infty) \end{cases},\ H \neq 0\right\}.$$

And

$$S_2 = \left\{ H : \mathrm{R} \to \mathrm{R} : \exists\, s_1, s_2 \in \mathrm{R} \text{ with } H(s) \begin{cases} \geq 0, s \in (-\infty, s_1) \\ \leq 0, s \in (s_1, s_2) \\ \geq 0, s \in (s_2, +\infty) \end{cases}, H \neq 0 \right\}.$$

The second order stochastic dominance depends on the values taken by the multiple m, the historical volatility σ and the implied volatility σ_i used to price the Call for the OBPI strategy. The determination of the second order stochastic dominance requires understanding the behavior of the function $H(x) = F_{V_T^{OBPI}}(x) - F_{V_T^{CBPI}}(x)$ based on the values taken by the multiple m. If $m = 1$, then the function H is strictly decreasing and presents a single point of intersection with the horizontal axis, thus $H \in S_1$. Therefore, for $m = 1$, $\sigma \leq \sigma_i$, we can conclude,

using theorem of Mosler [12] , that, if $H(x) \in S_1$ and $E\left[V_T^{OBPI}\right] \leq E\left[V_T^{CPPI}\right]$, then the CPPI strategy stochasti- cally dominates at the second order the OBPI strategy.

Theorem 3.2. Let $m = 1$ and $Call(S_0, K, r, \sigma_i) \geq Call(S_0, K, \mu, \sigma)$. Additionally, assume that $V_0 > Call(S_0, K, r, \sigma_i)$. Then, we deduce:

$$V_T^{CPPI} \succ_2 V_T^{OBPI}.$$

Proof. See Appendix A1.

Remark 3.1. Condition $V_0 > Call(S_0, K, r, \sigma_i)$ insures that $q > \alpha_T$ which allows proving the previous theorem. When $q = 1$ (as in Zagst and Kraus [10]), this condition is necessary satisfied.

3.1.2. The Third Order Stochastic Dominance

As mentioned by Zagst and Kraus [10] , the third order stochastic dominance (denoted by $\succ_3$) can be deduced under some specific assumptions.

Theorem 3.3. (Karlin-Novikov; Mosler [12])

Let V, V^* be non-negative random variables with finite second moments.

Denote $H(x) = F_V(x) - F_{V^*}(x)$ for all $x \in \mathrm{R}$. Then:

$$H \in S_2, E[V] \leq E[V^*], E[V^{*2}] \leq E[V^2] \Rightarrow V^* \succ_3 V.$$

The validation of the third order stochastic dominance requires the analysis of the condition $E\left[V^{*2}\right]\leq E\left[V^{2}\right]$ of previous Karlin and Novikov theorem. We get:

Theorem 3.4. Assuming that $\left(V_0 e^{rT}\right)^2 \leq E\left[V_T^{OBPI2}\right]$, there exists a value $m_{\max}$ of the multiple such that:

$$E\left[\left(V_T^{CPPI}\right)^2\right]\leq E\left[\left(V_T^{OBPI}\right)^2\right]\Leftrightarrow m\leq m_{\max}.$$

Proof. See Appendix A.2.

Using previous theorems, we deduce:

Theorem 3.5. Let $m_{\min}^1$ defined by:

$$m_{\min}^1 = 1+\frac{1}{(\mu-r)T}\ln\left(\frac{Call\left(S_0,\mu,\sigma\right)}{Call\left(S_0,r,\sigma_i\right)}\right)$$

and $m_{\min}=\max\left\{1,m_{\min}^1\right\}$.

Then, we get:

$$m\in\left[m_{\min},m_{\max}\right]\Rightarrow V_T^{CPPI}\succ_3 V_T^{OBPI}.$$

Proof. Condition $m\geq m_{\min}^1$ implies that $E\left[V_T^{CPPI}\right]\geq E\left[V_T^{OBPI}\right]$ (see Appendix A.1) while condition m ≤ mmax implies that $E\left[V_T^{CPPI2}\right]\leq E\left[V_T^{OBPI2}\right]$. Therefore, using Karlin and Novikov theorem, we deduce the result.

To illustrate these theoretical results, we consider the following numerical example: $\mu=7\%$, $\sigma=15\%$, $\sigma_i=18\%$, $r=3\%$, $T=5$ years, $V_0=S_0=100$, and $p=100\%$. Applying Relation (11), we deduce that $K\simeq 116$ and $q\simeq 0.86$. Table 1 illustrates the results of the third degree stochastic dominance for different values of the multiple $m=1,\cdots,5$.

Results of Table 1 show third order stochastic dominance of the CPPI strategy for $m=3\in[2.99,3.12]$.

Recall that, if the multiplier $m\geq m_{\min}=2.99$, we have $E\left[V_T^{CPPI}\right]\geq E\left[V_T^{OBPI}\right]$. However, for m ≥ mmax = 3.12,

we have $E\left[V_T^{CPPI2}\right]\geq E\left[V_T^{OBPI2}\right]$ and the sufficient condition of Karlin and Novikov theorem is no longer satisfied. The range of the multiple values, for which a stochastic dominance at the third order is verified, depends notably on the values of the implied volatility, the empirical volatility and the drift. Figures 1-3 illustrate this dependence.

Table 1. The third order stochastic dominance for multipliers equal to 1, ⋯, 5.

	$m=1$	$m=2$	$m=3$	$m=4$	$m=5$
m_{min}			2.99		
$m_{min} \prec m$	-	-	*	*	*
Condition S_2	-	*	*	*	*
m_{max}			3.12		
$m \prec m_{max}$	-	*	*	-	-
Third order SD if $m \in [m_{min}, m_{max}]$	-	-	*	-	-

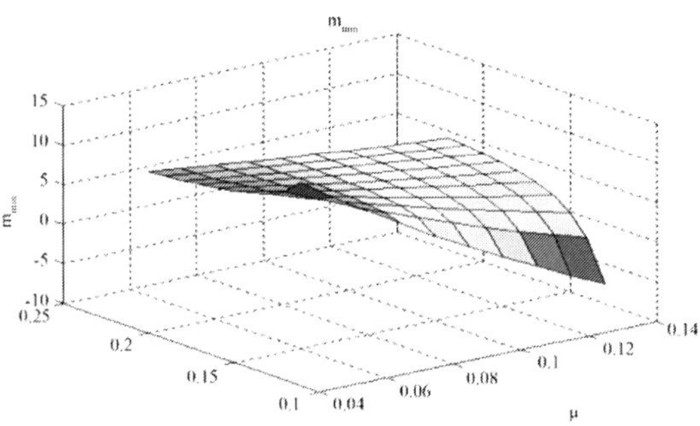

Figure 1. The value of mmin depending on the drift and the implied volatility.

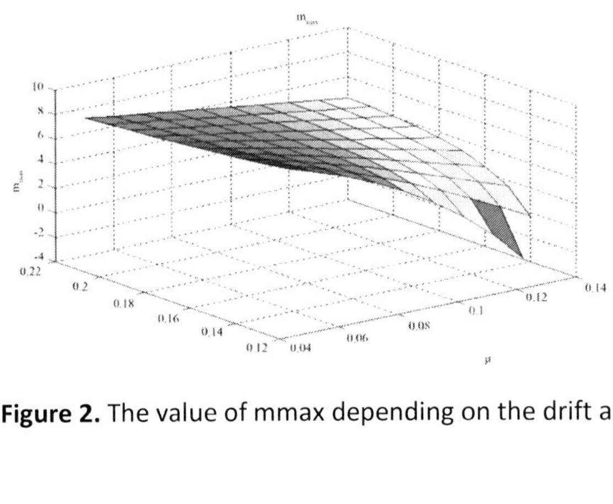

Figure 2. The value of mmax depending on the drift and the implied volatility.

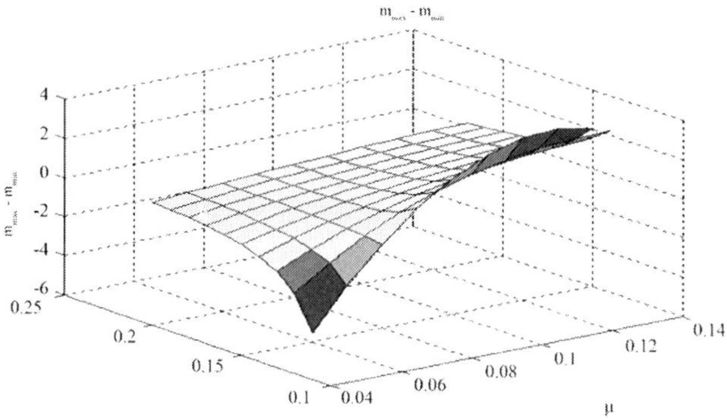

Figure 3. Difference of the upper and lower bounds on the multiplier.

The value of the lower bound $m_{\min}$ is a decreasing function with respect to the value of the drift μ. Indeed, when the drift increases, the expectation of the CPPI portfolio value increases more than that of the OBPI portfolio since the CPPI strategy is more allocated on the risky asset. The value of the lower bound $m_{\min}$ is not always an increasing function of the implied volatility.

The value of the upper bound $m_{\max}$ is a decreasing function with respect to the value of the drift μ. Indeed, when the drift increases, the expectation of the square of the CPPI portfolio value increases more than that of the OBPI portfolio since the CPPI strategy is more allocated on the risky asset. Therefore, the condition

$$E\left[V_T^{CPPI\,2}\right] \leq E\left[V_T^{OBPI\,2}\right]$$

is more stringent when the multiple m increases. The value of the lower bound $m_{\min}$

is almost always an increasing function of the implied volatility.

Previous stochastic dominance results have been established for the standard cases, i.e. the strategies are not capped. To deal with capped strategies as defined in Remarks (Capped CPPI) and (Capped OBPI), we have to conduct a numerical analysis. In a first step, we simulate the portfolios values using standard Monte Carlo methods; in a second step, we test the stochastic dominance properties.

4. STOCHASTIC DOMINANCE OF PORTFOLIO INSURANCE STRATEGIES

4.1. Stochastic Dominance: Numerical and Empirical Tests

In the empirical framework, the stochastic dominance has been pioneered for example by Kroll and Levy [13] . To avoid sampling errors due to i.i.d. assumptions, general stochastic dominance tests have been developed (e.g. Davidson and Duclos [14] ; Barrett and Donald [1] ; Post [15] ; Linton et al. [16] ; Scaillet and Topaloglou [17]). The tests introduced by Barrett and Donald [1] and Linton et al. [16] are based on a comparison of the cumulative density functions of studied perspectives. They are based on the Kolmogorov-Smirnov type tests. Barrett and Donald [1] examine the application of tests for any predetermined orders of stochastic dominance, SD_j, using several simulation and bootstrap methods to estimate an asymptotic p-value.

4.1.1. Stochastic Dominance and Hypothesis Formulation

Due to the characterizations of stochastic dominance, it is convenient to represent the various orders of stochastic dominance using the integral operators, $\mathcal{J}_j(\cdot;G)$, corresponding to successive integrations of the cumulative distribution function G to order $j-1$, namely:

$$\mathcal{J}_1(z;G)=G(z),$$

$$\mathcal{J}_2(z;G)=\int_0^z G(t)\,\mathrm{d}t=\int_0^z \mathcal{J}_1(t;G)\,\mathrm{d}t,$$

$$\mathcal{J}_3(z;G)=\int_0^z\int_0^s G(t)\,\mathrm{d}t=\int_0^z \mathcal{J}_2(s;G)\,\mathrm{d}s,$$

and so on.

The general hypotheses for testing stochastic dominance of G with respect to F at order j can be written compactly as:

$$H_0^j:\mathcal{J}_j(z;G)\le\mathcal{J}_j(z;F)\ \text{ for all } z\in(\underline{z},\overline{z}),$$

$$H_1^j:\mathcal{J}_j(z;G)\succ\mathcal{J}_j(z;F)\ \text{ for at least one } z\in(\underline{z},\overline{z}).$$

4.1.2. Test Statistics and Asymptotic Properties

In this paper, we test for stochastic dominance using the empirical distribution functions estimated from simulation of the two insurance portfolio strategies. The test of Linton et al. [16] allows for dependence in the data, and can be conducted with a limited number of assumptions. Suppose two prospects X and Y. Let N be the number of the realizations for the two prospects $\{X_i; i=1,\cdots,N\}$ and $\{Y_i; i=1,\cdots,N\}$. The null hypothesis is that a particular prospect X dominates the other one.

The empirical distributions used to construct the tests are respectively given by:

$$\widehat{F}_N(z)=\frac{1}{N}\sum_{i=1}^{N}1(X_i\le z),\widehat{G}_N(z)=\frac{1}{N}\sum_{i=1}^{N}1(Y_i\le z),$$

where j denotes the order of dominance and $1(.)$ denotes the indicator function.

The statistical test T_{Nj} for the full sample is defined by:

$$T_{Nj}=\sqrt{N}\sup_z\left(\mathcal{T}_j\left(z;\widehat{G}_N\right)-\mathcal{T}_j\left(z;\widehat{F}_N\right)\right).$$

The linear operator F_j is written as:

$$\mathcal{T}_j\left(z;\widehat{F}_N\right)=\frac{1}{N}\sum_{i=1}^{N}\mathcal{T}_j\left(z;1_{X_i}\right)=\frac{1}{N}\sum_{i=1}^{N}\frac{1}{(j-1)!}1(X_i\le z)(z-X_i)^{j-1}.$$

The second term of the linear operator is derived from Davidson and Duclos [14] .

We have also to define a method in order to obtain the critical value of the test. The standard bootstrap does not work because we need to impose the null hypothesis in that case, which is difficult because it is defined by a complicated system of inequalities. According to Linton et al. [16] , we apply the sub sampling method which is very simple to define and yet provide consistent critical values. Following the circular block method of Kläver [18] , we have to compute again the test statistic for the sub sample of size b for each of the $N-b+1$ different subsamples $\{W_i,\cdots,W_{(i+b-1)}\}$, where $W_i=\{X_{ki};k=1,2\}$ and $i=1,\cdots,N-b+1$, and for the subsamples $\{W_i,\cdots,W_N,W_1,\cdots,W_{(i+b-N-1)}\}$ where $i=N-b+2,\cdots,N$. Let $t_{(N,b,i)}$ be equal to the statistic t_b evaluated at the subsample $\{W_i,\cdots,W_{(i+b-1)}\}$ of size b. We have:

$$t_{(N,b,i)} = t_b\left(W_i,\cdots,W_{(i+b-1)}\right) \text{ for } i = 1,\cdots,N-b+1,$$

with

$$t_{(N,b,j)} = \sqrt{b}\sup_z\left(\mathcal{T}_j\left(z.;\widehat{G}_b\right) - \mathcal{T}_j\left(z;\widehat{F}_b\right)\right).$$

The underlying rationale is that one can approximate the sampling distribution of T_N using the distribution of the values of $t_{(N,b,j)}$ computed over $N-b+1$ different subsamples of size b, when $b/N \to 0$ and $b \to \infty$ as $N \to \infty$.

We consider that each of these sub samples is also a sample of the true sampling distribution of the original data.

Following Kläver [6] , we consider a sub sample size $b(N) = 10\sqrt{N}$.

Let $\widehat{p}_j$ denote the empirical p-value:

$$\widehat{p}_j = \frac{1}{N}\sum_{i=1}^{N} 1\left(t_{(N,b,j)} > T_{Nj}\right).$$

For $j = 1,2,3$, we reject the null hypothesis at α significance according to the following rule:

If $\widehat{p}_N^{(j)}(k) \leq \alpha$, we reject the null hypothesis of j-order stochastic dominance of variable X with respect to the variable Y.

If $\widehat{p}_N^{(j)}(k) \succ \alpha$, the variable X stochastically dominates the variable Y at the j-order.

4.1.3. Numerical Illustrations

In this subsection, we apply the tests of stochastic dominance in particular to check if the interval $[m_{\min}, m_{\max}]$ provided for the third order stochastic dominance between the CPPI strategies and OBPI strategies in previous theoretical subsection can be enlarged. We consider the case of a guarantee equal to 100% of the initially invested amount. Our numerical base case corresponds to a drift equal to 4.5%, an investment horizon equal to 8 years, an historical volatility equal to 15%. Our goal is to determine an order of stochastic dominance between the two insured portfolios by varying the multiplier of the

CPPI strategy into the interval $[2,9]$. We begin by varying the implicit volatility in $[20\%;32\%]$ as illustrated in Table 2.

We note that, for all cases in which the implied volatility far exceeds the historical volatility, the CPPI strategy, with a multiplier equal to 2, dominates the OBPI one.

We can also study the effect of the drift on the third order stochastic dominance (values of drift $\mu \in [4.5\%,7\%]$). We still fix the investment maturity to 8 years and consider a historical volatility equal to15%, a multiplier range in $[2,9]$, an implied volatility in $[27\%,32\%]$ and a guarantee level equal to 100%.

As shown in Table 3, the TSD is never verified, even if $m=2$ and $\sigma_i \in [27\%,32\%]$ when $\mu \geq 5\%$. We note that the CPPI strategy loses its attractiveness. Since $m=2$, we conclude that the CPPI strategy takes less advantage from the trend increase.

For lower trend levels and implicit volatility σ_i higher than the historical one σ, we get results given in Table 4.

Table 2. Third order stochastic dominance according to implicit volatility.

	m	μ	σ_i	σ	p-value	SD
$\sigma_i \leq \sigma$	2 to 9	4.5%		15%		NTSD
$\sigma_i > \sigma$	2	4.5%	27%	15%	0.0187	TSD
	[3.9]	4.5%	27%	15%		NTSD
	2	4.5%	30%	15%	0.2990	TSD
	[3.9]	4.5%	30%	15%		NTSD

Table 3. No third order stochastic dominance cases.

	m	μ	σ_i	σ	p-value	SD
$\sigma_i < \sigma$	[2.9]	[4.5%, 7%]	[11%.15%]	15%	0	NTSD
$\sigma_i > \sigma$	[2.9]	[4.5%, 7%]	[17%.21%]	15%		NTSD
$\sigma_i > \sigma$	[2.9]	[4.5%, 7%]	[21%.26%]	15%		NTSD
$\sigma_i > \sigma$	[2.9]	[5%, 7%]	[27%.32%]	15%		NTSD

Table 4. Third order stochastic dominance (low trend).

m	μ	σ_i	σ	p-value	SD
4	1%	17%	15%	0.0305	TSD
[2,→,5]	1%	17%	15%	0	NTSD
[2,→,5]	1%	[18% → 20%]	15%	0	NTSD
3	2%	18%	15%	0.3408	TSD
3	2%	19%	15%	0.0190	TSD
[2,→,5]	2%	[17% → 20%]	15%	0	NTSD
[2,→,5]	2%	[18%,19%]	15%	0	NTSD
2	3%	18%	15%	0.0533	TSD
2	3%	19%	15%	0.3913	TSD
[2,→,5]	[3%,→,4%]	[17% → 20%]	15%	0	NTSD

Remark 3.2. To summarize the numerical results:

-We have found that the CPPI method third order stochastically dominates the OBPI one for high implied volatility relatively to the empirical volatility;

-When the interval $[m_{\min}, m_{\max}]$ degenerates, we can find multiples for which the CPPI is stochastically dominated at the third order by OBPI;

-The implied volatility interval where the dominance relation is insured is larger for high values of implied volatility, for low values of the drift and for high values of the multiple.

-The TSD property of the CPPI strategy is rejected for the low values of σ_i with respect to σ.

-Through this numerical study, we can detect cases of third order stochastic dominance beyond the theoretical cases.

-Finally, when strategies are capped, the TSD property is generally not satisfied4.

5. CONCLUSION

In the present paper, we have compared the CPPI and OBPI strategies, mainly with respect to the third stochastic dominance (TSD). We find that the CPPI method third order stochastically dominates the OBPI one for high implied volatility relatively to the empirical volatility. We have checked the TSD of the CPPI method compared to the OBPI method for low values of the drift weighted by high values of the multiplier. We have shown that the relation of SDT is rejected for the low values of the implicit volatility with respect to the statistical one. Further extensions could be based on the use of almost stochastic dominance as defined by Leshno and Levy [19] , in order to extend the range of the multiple for which the CPPI dominates the OBPI.

REFERENCES

1. Barrett, G.F. and Donald, S.G. (2003) Consistent Tests for Stochastic Dominance. Econometrica, 71, 71-104.

2. Leland, H.E. and Rubinstein, M. (1976) The Evolution of Portfolio Insurance. In: Luskin, D.L., Ed., Portfolio Insurance: A Guide to Dynamic Hedging, Wiley.

3. Perold, A. (1986) Constant Portfolio Insurance. Working Paper, Harvard Business School.

4. Levy, H. (1992) Stochastic Dominance and Expected Utility: Survey and Analysis. Management Science, 38, 555-593. http://dx.doi.org/10.1287/mnsc.38.4.555

5. Levy, H. (2015) Stochastic Dominance: Investment Decision Making under Uncertainty. 3rd Edition, Springer-Verlag.

6. De Giorgi, E. (2005) Reward-Risk Portfolio Selection and Stochastic Dominance. Journal of Banking and Finance, 29, 895-926.

7. Darsinos, T. and Satchell, S. (2004) Generalizing Universal Performance Measures. Risk, 17, 80-84.

8. Bertrand, P. and Prigent, J.-L. (2005) Portfolio Insurance Strategies: OBPI versus CPPI. Finance, 26, 5-32.

9. Prigent, J.-L. (2007) Portfolio Optimization and Performance Analysis. Chapman & Hall, Boca Raton.

10. Zagst, R. and Kraus, J. (2011) Stochastic Dominance of Portfolio Insurance Strategies: OBPI versus CPPI. Annals of Operations Research, 185, 75-103. http://dx.doi.org/10.1007/s10479-009-0549-9

11. Van Der Vaart, A.W. and Wellner, J.A. (1996) Weak Convergence and Empirical Processes with Applications to Statistics. Springer-Verlag, New York.

12. Mosler, K.-C. (1982) Entscheidungsregeln bei Risiko-Multivariate stochastische Dominanz. Springer, Heidelberg.

13. Kroll, Y. and Levy, H. (1980) Sampling Errors and Portfolio Efficient Analysis. Journal of Financial and Quantitative Analysis, 15, 655-688.

14. Davidson, R. and Duclos, J.-Y. (2000) Statistical Inference for Stochastic Dominance and for the Measurement of Poverty and Inequality. Econometrica, 68, 1435-1464.

15. Post, T. (2003) Empirical Tests for Stochastic Dominance. Journal of Finance, 58, 1905-1932.

16. Linton, O., Maasoumi, E. and Whang, Y.-J. (2005) Consistent Testing for Stochastic Dominance under General Sampling Schemes. Review of Economic Studies, 72, 735-765.

17. Scaillet, O. and Topaloglou, N. (2010) Testing for Stochastic Dominance Efficiency. Journal of Business & Economic Statistics, 28, 169-180.

18. Kl&aumlver, H. (2005) Testing for Stochastic Dominance Using Circular Block Methods. Graduate School of Risk Management, University of K&oumlln.

19. Leshno, M. and Levy, H. (2002) Preferred by All and Preferred by Most Decision Makers: Almost Stochastic Dominance. Management Science, 48, 1074-1085.

Chapter 11

Construction of Zero Autocorrelation Stochastic Waveforms

Somantika Datta

Department of Mathematics, University of Idaho, Moscow, USA

ABSTRACT

Stochastic waveforms are constructed whose expected autocorrelation can be made arbitrarily small outside the origin. These waveforms are unimodular and complex-valued. Waveforms with such spike like autocorrelation are desirable in waveform design and are particularly useful in areas of radar and communications. Both discrete and continuous waveforms with low expected autocorrelation are constructed. Further, in the discrete case, frames for $\mathbb{C}^d$ are constructed from these waveforms and the frame properties of such frames are studied.

KEYWORDS

Autocorrelation; Frames; Stochastic Waveforms

1. INTRODUCTION

1.1. Motivation

Designing unimodular waveforms with an impulse-like autocorrelation is central in the general area of waveform design, and it is particularly relevant in several applications in the areas of radar and communications. In the former, the waveforms can play a role in effective target recognition, e.g., [1-8]; and in the latter they are used to address synchronization issues in cellular (phone) access technologies, especially code division multiple access (CDMA), e.g., [9,10]. The radar and communications methods combine in recent advanced multifunction RF systems (AMRFS). In radar there are two main reasons that the waveforms should be unimodular, that is, have constant amplitude. First, a transmitter can operate at peak power if the signal has constant peak amplitude—the system does not have to deal with the surprise of greater than expected amplitudes. Second, amplitude variations during transmission due to additive noise can be theoretically eliminated. The zero autocorrelation property ensures minimum interference between signals sharing the same channel.

Constructing unimodular waveforms with zero autocorrelation can be related to fundamental questions in harmonic analysis as follows. Let $\mathbb{R}$ be the real numbers, $\mathbb{Z}$ the integers, $\mathbb{C}$ the complex numbers, and set $\mathbb{T} = \mathbb{R}/\mathbb{Z}$. The aperiodic autocorrelation $A_X : \mathbb{Z} \to \mathbb{C}$ of a waveform $X : \mathbb{Z} \to \mathbb{C}$, is defined as

$$\forall k \in \mathbb{Z},\ A_X\left[k\right] = \lim_{N \to \infty} \frac{1}{2N+1} \sum_{m=-N}^{N} X\left[k+m\right] \overline{X\left[m\right]}. \tag{1}$$

A general problem is to characterize the family of positive bounded Radon measures F, whose inverse Fourier transforms are the autocorrelations of bounded waveforms X. A special case is when $F \equiv 1$ on $\mathbb{T}$ and X is unimodular on $\mathbb{Z}$. This is the same as when the autocorrelation of X vanishes except at 0, where it takes the value 1. In this case, X is said to have perfect autocorrelation. An extensive discussion on the construction of different classes of deterministic waveforms with perfect autocorrelation can be found in [11]. Instead of aperiodic waveforms that are defined on $\mathbb{Z}$, in some applications, it might be useful to construct periodic waveforms with similar vanishing properties of the autocorrelation function. Let $n \geq 1$ be an integer and $\mathbb{Z}_n$ be the finite group $\{0,1,\cdots,n-1\}$ with addition modulo n. The periodic autocorrelation $A_X : \mathbb{Z}_n \to \mathbb{C}$ of a waveform $X : \mathbb{Z}_n \to \mathbb{C}$ is defined as

$$\forall k = 0,1,2,\cdots,n-1,\ A_X\left[k\right] = \frac{1}{n} \sum_{m=0}^{n-1} X\left[m+k\right] \overline{X\left[m\right]}. \tag{2}$$

It is said that $X : \mathbb{Z}_n \to \mathbb{C}$ is a constant amplitude zero autocorrelation (CAZAC) waveform if each $|X[k]| = 1$ and

$$\forall k = 1,2,\cdots,n-1,\ A_X\left[k\right] = \frac{1}{n} \sum_{m=0}^{n-1} X\left[m+k\right] \overline{X\left[m\right]} = 0.$$

The literature on CAZACs is overwhelming. A good reference on this topic is [3], among many others. Literature on the general area of waveform design include [12-14]. Comparison between periodic and aperiodic autocorrelation can be found in [15].

Here the focus is on the construction of stochastic aperiodic waveforms. Henceforth, the reference to waveforms shall imply aperiodic waveforms unless stated otherwise. These waveforms are stochastic in nature and are constructed from certain random variables. Due to the stochastic nature of the construction, the expected value of the corresponding autocorrelation function is analyzed. It is desired that everywhere away from zero, the expectation of the autocorrelation can be made arbitrarily small. Such waveforms will be said to have almost perfect autocorrelation and will be called zero autocorrelation stochastic waveforms. First discrete waveforms, $X:\mathbb{Z}\to\mathbb{C}$, are constructed such that X has almost perfect autocorrelation and for all $n\in\mathbb{Z}$, $|X[n]|=1$. This approach is extended to the construction of continuous waveforms, $x:\mathbb{R}\to\mathbb{C}$, with similar spike like behavior of the expected autocorrelation and $|x(t)|=1$ for all $t\in\mathbb{R}$. Thus, these waveforms are unimodular. The stochastic and non-repetitive nature of these waveforms means that they cannot be easily intercepted or detected by an adversary. Previous work on the use of stochastic waveforms in radar can be found in [16-18], where the waveforms are only real-valued and not unimodular. In comparison, the waveforms constructed here are complex valued and unimodular. In addition, frame properties of frames constructed from these stochastic waveforms are discussed. This is motivated by the fact that frames have become a standard tool in signal processing. Previously, a mathematical characterization of CAZACs in terms of finite unit-normed tight frames (FUNTFs) has been done in [2].

1.2. Notation and Mathematical Background

Let X be a random variable with probability density function f. Assuming X to be absolutely continuous, the expectation of X denoted by $E(X)$, is

$$E(X)=\int_{\mathbb{R}} xf(x)\mathrm{d}x.$$

The Gaussian random variable has probability density function given by $f(x)=\frac{1}{\sigma\sqrt{2\pi}}e^{-\frac{1}{2}\left(\frac{x-\mu}{\sigma}\right)^2}$. The mean or expectation of this random variable is μ and the variance, $V(X)$, is σ^2. In this case it is also said that X follows a normal distribution and is written as $X\sim N(\mu,\sigma^2)$. The characteristic function of X at t,

$E\left(e^{itX}\right)$, is denoted by $\phi_X(t)$. For further properties of expectation and characteristic function of a random variable the reader is referred to [19].

Let $\mathcal{H}$ be a Hilbert space and let $V=\{v_k, k\in\mathcal{K}\}$, where $\mathcal{K}$ is some index set, be a collection of vectors in $\mathcal{H}$. Then V is said to be a frame for $\mathcal{H}$ if there exist constants A and B, $0<A\le B<\infty$, such that for any $v\in\mathcal{H}$

$$A\|v\|^2 \le \sum_{k\in\mathcal{K}} \left|\langle v, v_k\rangle\right|^2 \le B\|v\|^2.$$

The constants A and B are called the frame bounds. Thus a frame can be thought of as a redundant basis. In fact, for a finite dimensional vector space, a frame is the same as a spanning set. If $A=B$, the frame is said to be tight. Orthonormal bases are special cases of tight frames and for these, $A=B=1$.

If V is a frame for $\mathcal{H}$ then the map $F:\mathcal{H}\to\ell_2(\mathcal{K})$ given by $F(v)=\{\langle v,v_k\rangle : k\in\mathcal{K}\}$ is called the analysis operator. The synthesis operator is the adjoint map $F^*:\ell_2(\mathcal{K})\to\mathcal{H}$, given by

$$F^*\left(\{a_k\}\right)=\sum_{k\in\mathcal{K}} a_k v_k.$$

The frame operator $\mathcal{F}:\mathcal{H}\to\mathcal{H}$ is given by $\mathcal{F}=F^*F$. For a tight frame, the frame operator is just a constant multiple of the identity, i.e., $\mathcal{F}=A\mathcal{I}$, where $\mathcal{I}$ is the identity map. Every $v\in\mathcal{H}$ can be represented as

$$v=\sum_{k\in\mathcal{K}}\left\langle v,\mathcal{F}^{-1}v_k\right\rangle v_k=\sum_{k\in\mathcal{K}}\left\langle v,v_k\right\rangle \mathcal{F}^{-1}v_k.$$

Here $\{\mathcal{F}^{-1}v_k\}$ is also a frame and is called the dual frame. For a tight frame, $\mathcal{F}^{-1}$ is just $\frac{1}{A}\mathcal{F}$. Tight frames are thus highly desirable since they offer a computationally simple reconstruction formula that does not involve inverting the frame operator. The minimum and maximum eigenvalues of $\mathcal{F}$ are the optimal lower and upper frame bounds respectively [20]. Thus, for a tight frame all the eigenvalues of the frame operator are equal to each other. For the general theory on frames one can refer to [20,21].

1.3. Outline

The construction of discrete unimodular stochastic waveforms, $X:\mathbb{Z}\to\mathbb{C}$, with almost perfect autocorrelation is done in Section 2. This is first done with the Gaussian random variable and then generalized to other random variables. The variance of the autocorrelation is also estimated. The section also addresses the construction of stochastic waveforms in higher dimensions, i.e., construction of $v:\mathbb{Z}\to\mathbb{C}^d$, that have almost perfect autocorrelation and are unit-normed, considering the usual norm in $\mathbb{C}^d$. In Section 3 the construction of unimodular continuous waveforms with almost perfect autocorrelation is done using Brownian motion.

As mentioned in Section 1.2, frames are now a standard tool in signal processing due to their effectiveness in robust signal transmission and reconstruction. In Section 4, frames in $\mathbb{C}^d\,(d\geq 2)$ are constructed from the discrete waveforms of Section 2 and the nature of these frames is analyzed. In particular, the maximum and minimum eigenvalues of the frame operator are estimated. This helps one to understand how close these frames are to being tight. Besides, it follows, from the eigenvalue estimates, that the matrix of the analysis operator, F, for such frames, can be used as a sensing matrix in compressed sensing.

2. CONSTRUCTION OF DISCRETE STOCHASTIC WAVEFORMS

In this section discrete unimodular waveforms, $X:\mathbb{Z}\to\mathbb{C}$, are constructed from random variables such that the expectation of the autocorrelation can be made arbitrarily small everywhere except at the origin. First, such a construction is done using the Gaussian random variable. Next, a general characterization of all random variables that can be used for the purpose is given.

2.1. Construction from Gaussian Random Variables

Let $\{Y_\ell\}_{\ell\in\mathbb{Z}}$ be independent identically distributed (i.i.d.)

random variables following a Gaussian or normal distribution with mean 0 and variance σ^2, i.e., $Y_\ell \sim N(0,\sigma^2)$. Define $X:\mathbb{Z}\to\mathbb{C}$ by

$$\forall n\in\mathbb{Z},\quad X[n]=e^{\frac{2\pi i}{\varepsilon}\sum_{\ell=-n}^{n} Y_\ell} \tag{3}$$

where i is $\sqrt{-1}$. Thus, for each n, $|X[n]|=1$ and X is unimodular. The autocorrelation of X at $k \in \mathbb{Z}$ is

$$A_X[k] = \lim_{N\to\infty} \frac{1}{2N+1} \sum_{n=-N}^{N} X[n+k]\overline{X[n]}$$

where the limit is in the sense of probability. Theorem 2.1 shows that the waveform given by (3) has autocorrelation whose expectation can be made arbitrarily small for all integers $k \neq 0$.

Theorem 2.1. Given $\varepsilon > 0$, the waveform $X:\mathbb{Z}\to\mathbb{C}$ defined in (3) has autocorrelation A_X such that

$$E(A_X[k]) = \begin{cases} 1 & \text{if } k = 0 \\ e^{-|k|\sigma^2\left(\frac{2\pi}{\varepsilon}\right)^2} & \text{if } k \neq 0. \end{cases}$$

Proof. 1) When $k = 0$,

$$A_X[0] = \lim_{N\to\infty} \frac{1}{2N+1} \sum_{n=-N}^{N} X[n]\overline{X[n]} = 1.$$

and so $E(A_X[0]) = 1$.

2) Let $k > 0$. One would like to calculate

$$E(A_X[k]) = E\left(\lim_{N\to\infty} \frac{1}{2N+1} \sum_{n=-N}^{N} X[n+k]\overline{X[n]}\right).$$

Let $g_N(X) = \frac{1}{2N+1}\sum_{n=-N}^{N} X[n+k]\overline{X[n]}$. Then $|g_N(X)| \leq 1$. Let $h(X)=1$. Then for each N,

$|g_N(X)| \leq h(X)$ and $E(h(X))=1$. Thus, by the Dominated Convergence Theorem [19], which justifies the interchange of limit and integration below, one obtains

$$\begin{aligned}
E\left(A_X[k]\right) &= E\left(\lim_{N\to\infty}\frac{1}{2N+1}\sum_{n=-N}^{N} X[n+k]\overline{X[n]}\right). \\
&= \lim_{N\to\infty}\frac{1}{2N+1}\sum_{n=-N}^{N} E\left(X[n+k]\overline{X[n]}\right) \\
&= \lim_{N\to\infty}\frac{1}{2N+1}\sum_{n=-N}^{N} E\left(e^{\frac{2\pi i}{\varepsilon}\sum_{\ell=-n-k}^{n+k} Y_\ell}\, e^{-\frac{2\pi i}{\varepsilon}\sum_{m=-n}^{n} Y_m}\right) \\
&= \lim_{N\to\infty}\frac{1}{2N+1}\sum_{n=-N}^{N} E\left(e^{\frac{2\pi i}{\varepsilon}\left(\sum_{\ell=n+1}^{n+k} Y_\ell + \sum_{m=-n-k}^{-n-1} Y_m\right)}\right) \\
&= \lim_{N\to\infty}\frac{1}{2N+1}\sum_{n=-N}^{N}\left[E\left(e^{\frac{2\pi i}{\varepsilon}Y_1}\right)\right]^{2k} \\
&= \left[E\left(e^{\frac{2\pi i}{\varepsilon}Y_1}\right)\right]^{2k} = \left[\phi_{Y_1}\left(\frac{2\pi}{\varepsilon}\right)\right]^{2k}
\end{aligned}$$

where the last line uses the fact that the Y_ℓs are i.i.d. random variables. Here $\phi_{Y_1}\left(\frac{2\pi}{\varepsilon}\right)$ is the characteristic function at $\frac{2\pi}{\varepsilon}$ of Y_1 which is the same as that for any other Y_ℓ due to their identical distribution. The characteristic function at $\frac{2\pi}{\varepsilon}$ of a Gaussian random variable with mean 0 and variance σ^2 is $e^{-\frac{\sigma^2}{2}\left(\frac{2\pi}{\varepsilon}\right)^2}$. Thus

$$E\left(A_X[k]\right) = \left[e^{-\frac{\sigma^2}{2}\left(\frac{2\pi}{\varepsilon}\right)^2}\right]^{2k} = e^{-k\sigma^2\left(\frac{2\pi}{\varepsilon}\right)^2}.$$

3) When $k > 0$, a similar calculation for $A_X[-k]$ gives

$$E\left(A_X[-k]\right) = \left[e^{-\frac{\sigma^2}{2}\left(\frac{2\pi}{\varepsilon}\right)^2}\right]^{2k} = e^{-k\sigma^2\left(\frac{2\pi}{\varepsilon}\right)^2}.$$

Together, this shows that given ε and any $k \neq 0$,

$$E\left(A_X[k]\right) = e^{-|k|\sigma^2\left(\frac{2\pi}{\varepsilon}\right)^2}$$

which indicates that the expectation of the autocorrelation at any integer $k \neq 0$ can be made arbitrarily small depending on the choice of ε.

As shown in Theorem 2.1 the expectation of the autocorrelation can be made arbitrarily small but this is not useful unless one can estimate the variance of the autocorrelation. Denoting the variance of $A_X[k]$ by $V(A_X[k])$ one has

$$V(A_X[k]) = E\left(\left|A_X[k]\right|^2\right) - \left|E(A_X[k])\right|^2$$
$$= E\left(\left|A_X[k]\right|^2\right) - e^{-2|k|\sigma^2\left(\frac{2\pi}{\varepsilon}\right)^2}.$$

First consider $k > 0$;

$$\left|A_X[k]\right|^2 = \left(\lim_{N\to\infty}\frac{1}{2N+1}\sum_{n=-N}^{N} X[n+k]\overline{X[n]}\right)\left(\lim_{M\to\infty}\frac{1}{2M+1}\sum_{m=-M}^{M}\overline{X[m+k]}X[m]\right)$$
$$= \lim_{N\to\infty}\lim_{M\to\infty}\frac{1}{(2N+1)}\frac{1}{(2M+1)}\sum_{n=-N}^{N}\sum_{m=-M}^{M} X[n+k]\overline{X[n]}\,\overline{X[m+k]}X[m]$$
$$= \lim_{N\to\infty}\lim_{M\to\infty}\frac{1}{(2N+1)}\frac{1}{(2M+1)}\sum_{n=-N}^{N}\sum_{m=-M}^{M} e^{\frac{2\pi}{\varepsilon}i\left(\sum_{j=1}^{k}Y_{n-j}+\sum_{j=1}^{k}Y_{n+j}-\sum_{j=1}^{k}Y_{m-j}-\sum_{j=1}^{k}Y_{m+j}\right)}.$$

By applying the Lebesgue Dominated Convergence Theorem one can bring the expectation inside the double sum to get

$$E\left(\left|A_X[k]\right|^2\right) = \lim_{N\to\infty}\lim_{M\to\infty}\frac{1}{(2N+1)}\frac{1}{(2M+1)}\sum_{n=-N}^{N}\sum_{m=-M}^{M} E\left(e^{\frac{2\pi}{\varepsilon}i\left(\sum_{j=1}^{k}Y_{n-j}+\sum_{j=1}^{k}Y_{n+j}-\sum_{j=1}^{k}Y_{m-j}-\sum_{j=1}^{k}Y_{m+j}\right)}\right).$$

The sum

$$\sum_{j=1}^{k}Y_{n-j}+\sum_{j=1}^{k}Y_{n+j}-\sum_{j=1}^{k}Y_{m-j}-\sum_{j=1}^{k}Y_{m+j} \quad (4)$$

may have cancelations among terms involving n with terms involving m. Suppose that for a fixed n and m there are $\tilde{k}_{m,n}$ indices that cancel in each of the four sums in (4). Due to symmetry, the same number i.e., $\tilde{k}_{m,n}$ of terms will cancel in each sum. Depending on n and m, $\tilde{k}_{m,n}$ lies between 0 and k, i.e., $0 \leq \tilde{k}_{mn} \leq k$. For the sake of making the notation less cumbersome, $\tilde{k}_{m,n}$ will from now on be written as $\tilde{k}$. When $m = n$, $\tilde{k} = k$. If $m > n + k$ or $n > m + k$ then

$\tilde{k}=0$. Each sum in (4) has k terms and $\tilde{k}$ of these get cancelled leaving $(k-\tilde{k})$ terms. One can re-index the variables in (4) and write it as

$$\sum_{j=1}^{k} Y_{-n-j}+\sum_{j=1}^{k} Y_{n+j}-\sum_{j=1}^{k} Y_{-m-j}-\sum_{j=1}^{k} Y_{m+j}=\pm Y_{\ell_1}\pm\cdots\pm Y_{l_{4(k-\tilde{k})}}$$

where the sign depends on whether m is less than or greater than n. Thus

$$E\left(\left|A_X[k]\right|^2\right)=\lim_{N\to\infty}\lim_{M\to\infty}\frac{1}{(2N+1)}\frac{1}{(2M+1)}\sum_{n=-N}^{N}\sum_{m=-M}^{M}E\left(e^{\frac{2\pi}{\varepsilon}i\left(\pm Y_{\ell_1}\pm\cdots\pm Y_{l_{4(k-\tilde{k})}}\right)}\right).$$

Due to the independence of the Y_ℓ s, this means

$$E\left(\left|A_X[k]\right|^2\right)=\lim_{N\to\infty}\lim_{M\to\infty}\frac{1}{(2N+1)}\frac{1}{(2M+1)}\sum_{n=-N}^{N}\sum_{m=-M}^{M}\left[\phi_{Y_1}\left(\pm\frac{2\pi}{\varepsilon}\right)\right]^{4(k-\tilde{k})}$$

$$=\lim_{N\to\infty}\lim_{M\to\infty}\frac{1}{(2N+1)}\frac{1}{(2M+1)}\sum_{n=-N}^{N}\sum_{m=-M}^{M}e^{-\frac{\sigma^2}{2}\left(\frac{2\pi}{\varepsilon}\right)^2 4(k-\tilde{k})}.$$

The minimum is attained for $\tilde{k}=0$ and the maximum at $\tilde{k}=k$. Thus

$$E\left(\left|A_X[k]\right|^2\right)\le\lim_{N\to\infty}\lim_{M\to\infty}\frac{1}{(2N+1)}\frac{1}{(2M+1)}\sum_{n=-N}^{N}\sum_{m=-M}^{M}1=1$$

and

$$E\left(\left|A_X[k]\right|^2\right)\ge\lim_{N\to\infty}\lim_{M\to\infty}\frac{1}{(2N+1)}\frac{1}{(2M+1)}\sum_{n=-N}^{N}\sum_{m=-M}^{M}e^{-\frac{\sigma^2}{2}\left(\frac{2\pi}{\varepsilon}\right)^2 4k}=e^{-\frac{\sigma^2}{2}\left(\frac{2\pi}{\varepsilon}\right)^2 4k}.$$

This gives

$$0\le V\left(A_X[k]\right)\le 1-e^{-2k\sigma^2\left(\frac{2\pi}{\varepsilon}\right)^2}.$$

A similar calculation can be done for $k<0$. Thus for $k\neq 0$,

$$0\le V\left(A_X[k]\right)\le 1-e^{-2|k|\sigma^2\left(\frac{2\pi}{\varepsilon}\right)^2}.$$

2.2. Generalizing the Construction to Other Random Variables

So far the construction of discrete unimodular zero autocorrelation stochastic waveforms has been based on Gaussian random variables. This construction can be generalized to many other random variables. The unimodularity of the waveforms is not affected by using a different random variable. The following theorem characterizes the class of random variables that can be used to get the desired autocorrelation.

Theorem 2.2. Let $\{Y_\ell\}_{\ell\in\mathbb{Z}}$ be a sequence of i.i.d. random variables with characteristic function ϕ_Y. Suppose that the probability density function of the Y_ℓ s is even and that $\phi_Y(t)$ goes to 0 as t goes to infinity. Then, given ε, the waveform $X:\mathbb{Z}\to\mathbb{C}$ given by

$$X[n]=\mathrm{e}^{\frac{2\pi}{\varepsilon}i\sum_{\ell=-n}^{n}Y_\ell}$$

has almost perfect autocorrelation.

Proof. Since the density function of each Y_ℓ is even this means that the characteristic function is real valued [19]. Following the calculation in the proof of Theorem 2.1, the expected autocorrelation of X for $k\neq 0$ is

$$E(A_X[k])=\left[\phi_Y\left(\frac{2\pi}{\varepsilon}\right)\right]^{2|k|}$$

and this goes to zero with ε by the hypothesis. □

Example 2.3. Suppose the Y_ℓ s follow a bilateral distribution that has density $\mathrm{e}^{-|x|}$ with $x\in(-\infty,\infty)$ and characteristic function $\phi_Y(t)=\frac{1}{1+t^2}$. Then for $k\neq 0$,

$$E(A_X[k])=\left[\frac{1}{1+\left(\frac{2\pi}{\varepsilon}\right)^2}\right]^{2|k|}$$

and this can be made arbitrarily small with ε.

In the same way as was done in the Gaussian case, for $k>0$,

$$E\left(\left|A_X[k]\right|^2\right)$$
$$= \lim_{N\to\infty}\lim_{M\to\infty}\frac{1}{(2N+1)}\frac{1}{(2M+1)}\sum_{n=-N}^{N}\sum_{m=-M}^{M}\left[\phi_{Y_1}\left(\pm\frac{2\pi}{\varepsilon}\right)\right]^{4(k-\tilde{k})}$$
$$\le 1$$

and

$$E\left(\left|A_X[k]\right|^2\right) \ge \left[\frac{1}{1+\left(\frac{2\pi}{\varepsilon}\right)^2}\right]^{4k}.$$

Thus

$$0 \le V\left(A_X[k]\right) \le 1-\left(\frac{1}{1+\left(\frac{2\pi}{\varepsilon}\right)^2}\right)^{4|k|}.$$

Example 2.4. Suppose that the Y_ℓ s follow the Cauchy distribution with density function $\frac{1}{\pi(1+x^2)}$. Note thatdisregarding the constant π, this is the characteristic function of the random variable considered in Example 2.3. The characteristic function of the Y_ℓ s is now $e^{-|t|}$, the same as the distribution function in Example 2.3. For $k \neq 0$,

$$E\left(A_X[k]\right) = \left[\phi_{Y_1}\left(\frac{2\pi}{\varepsilon}\right)\right]^{2|k|} = e^{-\frac{4\pi|k|}{\varepsilon}}$$

which can be made arbitrarily small with ε. Also,

$$0 \le V\left(A_X[k]\right) \le 1-e^{-\frac{8\pi|k|}{\varepsilon}}.$$

2.3. Higher Dimensional Case

Here one is interested in constructing waveforms $v:\mathbb{Z}\to\mathbb{C}^d$, $d\geq 2$. It is desired that v has unit norm and the expectation of its autocorrelation can be made arbitrarily small. One way to construct v is based on the construction of the one dimensional example given in Section 2.1. This is motivated by the higher dimensional construction in the deterministic case [2]. As before, $\{Y_\ell\}_{\ell\in\mathbb{Z}}$ is a sequence of i.i.d. Gaussian random variables with mean zero and variance σ^2.

Next, one defines $X[n]=e^{\frac{2\pi}{\varepsilon}i\sum_{\ell=-n}^{n}Y_\ell}$. The waveform $v:\mathbb{Z}\to\mathbb{C}^d$ is then defined as

$$\forall m\in\mathbb{Z},\ v[m]=\frac{1}{\sqrt{d}}\begin{bmatrix} X[m] \\ X[m+1] \\ \vdots \\ X[m+d-1] \end{bmatrix}. \tag{5}$$

In this case, the autocorrelation is given by

$$A_v[k]=\lim_{N\to\infty}\frac{1}{2N+1}\sum_{n=-N}^{N}\langle v[n+k],v[n]\rangle \tag{6}$$

where $\langle\cdot,\cdot\rangle$ is the usual inner product in $\mathbb{C}^d$. The length or norm of any $v[m]$ is thus given by

$$\|v[m]\|^2=\langle v[m],v[m]\rangle.$$

From (5),

$$\|v[m]\|^2=\frac{1}{d}\sum_{n=0}^{d-1}X[m+n]\overline{X[m+n]}=\frac{d}{d}=1.$$

Thus the $v[m]$s are unit-normed. The following Theorem 2.5 shows that the expected autocorrelation of v can be made arbitrarily small everywhere except at the origin.

Theorem 2.5. Given $\varepsilon>0$, the waveform $v:\mathbb{Z}\to\mathbb{C}^d$ defined in (5) has autocorrelation A_v such that

$$E\left(A_v[k]\right)=\begin{cases}1 & \text{if } \ k=0 \\ e^{-|k|\sigma^2\left(\frac{2\pi}{\varepsilon}\right)^2} & \text{if } \ k\neq 0.\end{cases}$$

Proof. As defined in (6),

$$A_v[k]=\lim_{N\to\infty}\frac{1}{2N+1}\sum_{n=-N}^{N}\langle v[n+k],v[n]\rangle$$

When $k=0$,

$$A_v[0]=\lim_{N\to\infty}\frac{1}{2N+1}\sum_{n=-N}^{N}\left\|v[n]\right\|^2=1.$$

Thus,

$$E\left(A_v[0]\right)=1.$$

For $k\neq 0$, due to (5),

$$\langle v[n+k],v[n]\rangle=\frac{1}{d}\left\langle\begin{bmatrix}X[n+k]\\X[n+k+1]\\\vdots\\X[n+k+d-1]\end{bmatrix},\begin{bmatrix}X[n]\\X[n+1]\\\vdots\\X[n+d-1]\end{bmatrix}\right\rangle$$
$$=\frac{1}{d}\left(X[n+k]\overline{X[n]}+X[n+k+1]\overline{X[n+1]}+\cdots+X[n+k+d-1]\overline{X[n+d-1]}\right).$$

Consider $k>0$.

$$\begin{aligned}E\left(A_v[k]\right)&=\lim_{N\to\infty}\frac{1}{2N+1}\sum_{n=-N}^{N}E\left(\langle v[n+k],v[n]\rangle\right)\\&=\lim_{N\to\infty}\frac{1}{2N+1}\sum_{n=-N}^{N}E\left(\frac{1}{d}\sum_{m=0}^{d-1}X[n+k+m]\overline{X[n+m]}\right)\\&=\lim_{N\to\infty}\frac{1}{2N+1}\sum_{n=-N}^{N}\frac{1}{d}\sum_{m=0}^{d-1}E\left(X[n+k+m]\overline{X[n+m]}\right)\\&=\lim_{N\to\infty}\frac{1}{2N+1}\sum_{n=-N}^{N}\frac{1}{d}\sum_{m=0}^{d-1}E\left(e^{\frac{2\pi}{\varepsilon}i\left(Y_{-(n+m+k)}+\cdots+Y_{-(n+m+1)}+Y_{n+m+1}+\cdots+Y_{n+m+k}\right)}\right)\\&=\lim_{N\to\infty}\frac{1}{2N+1}\sum_{n=-N}^{N}\frac{1}{d}\sum_{m=0}^{d-1}E\left(e^{\frac{2\pi}{\varepsilon}iY_1}\right)^{2k}=e^{-\sigma^2k\left(\frac{2\pi}{\varepsilon}\right)^2}.\end{aligned}$$

Similarly, for $k<0$, one gets

$$E\left(A_v[k]\right)=\mathrm{e}^{\sigma^2 k\left(\frac{2\pi}{\varepsilon}\right)^2}.$$

Thus the waveform v as defined in this section is unit-normed and has autocorrelation that can be made arbitrarily small.

Remark 2.6. As in the one dimensional construction, it is easy to see that here too the construction can be done with random variables other than the Gaussian. In fact, all random variables that can be used in the one dimensional case, i.e., ones satisfying the properties of Theorem 2.2, can also be used for the higher dimensional construction.

2.4. Remark on the Periodic Case

It can be shown that the periodic case follows the same nature as the aperiodic case. The sequence $X:\mathbb{Z}_n\to\mathbb{C}$ is defined in the same way as in Section 2.1, i.e.,

$$\forall m\in\{0,1,\cdots,n-1\},\ X[m]=\mathrm{e}^{\frac{2\pi}{\varepsilon}i\sum_{\ell=-m}^{m}Y_\ell}$$

where $Y_\ell\sim N\left(0,\sigma^2\right)$. Following the definition given in (2), when $k=0$,

$$A_X[0]=\frac{1}{n}\sum_{m=0}^{n-1}X[m]\overline{X[m]}=1.$$

When $k\neq 0$, the expectation of the autocorrelation is

$$E\left(A_X[k]\right)=\frac{1}{n}\sum_{m=0}^{n-1}E\left(X[m+k]\overline{X[m]}\right).$$

For $k>0$,

$$E\left(A_X[k]\right)=\frac{1}{n}\sum_{m=0}^{n-1}E\left(\mathrm{e}^{\frac{2\pi i}{\varepsilon}\sum_{\ell=-m-k}^{m+k}Y_\ell}\mathrm{e}^{-\frac{2\pi i}{\varepsilon}\sum_{j=-m}^{m}Y_j}\right)$$

$$=\frac{1}{n}\sum_{m=0}^{n-1}E\left(\mathrm{e}^{\frac{2\pi i}{\varepsilon}\left(\sum_{\ell=m+1}^{m+k}Y_\ell+\sum_{j=-m-k}^{-m-1}Y_j\right)}\right)=\frac{1}{n}\sum_{m=0}^{n-1}\left[E\left(\mathrm{e}^{\frac{2\pi i}{\varepsilon}Y_1}\right)\right]^{2k}$$

$$=\left[E\left(\mathrm{e}^{\frac{2\pi i}{\varepsilon}Y_1}\right)\right]^{2k}=\left[\phi_{Y_1}\left(\frac{2\pi}{\varepsilon}\right)\right]^{2k}=\mathrm{e}^{-k\sigma^2\left(\frac{2\pi}{\varepsilon}\right)^2}$$

where one uses the fact that the Y_ℓ s are i.i.d.. A similar calculation for negative values of k suggests that the autocorrelation can be made arbitrarily small, depending on ε, for all non-zero values of k. Also, as in the aperiodic case, this result can be obtained for random variables other than the Gaussian.

3. CONSTRUCTION OF CONTINUOUS STOCHASTIC WAVEFORMS

In this section continuous waveforms with almost perfect autocorrelation are constructed from a one dimensional Brownian motion.

For a continuous waveform $x:\mathbb{R}\to\mathbb{C}$, the autocorrelation $A_x:\mathbb{R}\to\mathbb{C}$ can be defined as

$$A_x(s)=\lim_{T\to\infty}\frac{1}{2T}\int_{-T}^{T}x(t+s)\overline{x(t)}\,\mathrm{d}t. \tag{7}$$

Let $\{W(t);t>0\}$ be a one dimensional Brownian motion. Then $W(t)$ satisfies

- $W(0)=0$
- $W(t+s)-W(s)\sim N\left(0,\sigma^2 t\right)$
- $W(t_{i+1})-W(t_i)\,0<t_1<\cdots<t_k$, are independent random variables.

Theorem 3.1. Let $W(t)$ be the one dimensional Brownian motion and $\varepsilon>0$ be given. Define $x:\mathbb{R}\to\mathbb{C}$ by

$$x(t) = e^{\frac{2\pi}{\varepsilon} iW(t)} \text{ for } t \geq 0,$$

and $x(-t) = x(t)$. Then the autocorrelation of x, A_x, satisfies

$$E(A_x(s)) = \begin{cases} 1 & \text{if } s = 0 \\ e^{-\frac{\sigma^2}{2}|s|\left(\frac{2\pi}{\varepsilon}\right)^2} & \text{if } s \neq 0. \end{cases}$$

Proof. We would like to evaluate

$$E(A_x(s)) = E\left(\lim_{T\to\infty} \frac{1}{2T} \int_{-T}^{T} x(t+s)\overline{x(t)}\, \mathrm{d}t \right).$$

Let $s > 0$ and let $g_T(s) = \frac{1}{2T} \int_{-T}^{T} x(t+s)\overline{x(t)}\, \mathrm{d}t.$

$$\begin{aligned} E(g_T) &= \frac{1}{2T} \int_{-T}^{T} E\left(x(t+s)\overline{x(t)} \right) \mathrm{d}t \\ &= \frac{1}{2T} \int_{-T}^{T} E\left(e^{\frac{2\pi}{\varepsilon} i(W(t+s)-W(t))} \right) \mathrm{d}t \\ &= \frac{1}{2T} \int_{-T}^{T} \phi_{W(t+s)-W(t)}\left(\frac{2\pi}{\varepsilon} \right) = e^{-\frac{\sigma^2}{2}s\left(\frac{2\pi}{\varepsilon}\right)^2} < \infty. \end{aligned}$$

Thus each g_T is integrable and further $|g_T| \leq 1$. Let $h(t) = 1$; $t \in \mathbb{R}$. Then $E(h) = 1$. Therefore, by the Dominated Convergence Theorem, and properties of Brownian motion and characteristic functions, one gets

$$\begin{aligned} E(A_x(s)) &= E\left(\lim_{T\to\infty} \frac{1}{2T} \int_{-T}^{T} x(t+s)\overline{x(t)}\, \mathrm{d}t \right) \\ &= \lim_{T\to\infty} \frac{1}{2T} \int_{-T}^{T} E\left(e^{\frac{2\pi}{\varepsilon} i(W(t+s)-W(t))} \right) \mathrm{d}t \\ &= \lim_{T\to\infty} \frac{1}{2T} \int_{-T}^{T} \phi_{W(t+s)-W(t)}\left(\frac{2\pi}{\varepsilon} \right) = e^{-\frac{\sigma^2}{2}s\left(\frac{2\pi}{\varepsilon}\right)^2} \end{aligned}$$

which can be made arbitrarily small based on ε. Similarly,

$$E\left(A(-s)\right) = \lim_{T\to\infty}\frac{1}{2T}\int_{-T}^{T}\phi_{W(t)-W(t-s)}\left(-\frac{2\pi}{\varepsilon}\right)$$
$$= e^{-\frac{\sigma^2}{2}s\left(-\frac{2\pi}{\varepsilon}\right)^2} = e^{-\frac{\sigma^2}{2}s\left(\frac{2\pi}{\varepsilon}\right)^2}. \quad \square$$

4. CONNECTION TO FRAMES

Consider the mapping $v:\mathbb{Z}\to\mathbb{C}^d$ given by

$$v(k) = \frac{1}{\sqrt{d}}\begin{pmatrix} X[k] \\ X[k+1] \\ \vdots \\ X[k+d-1] \end{pmatrix} \tag{8}$$

where $X[k] = e^{\frac{2\pi}{\varepsilon}i\sum_{\ell=-k}^{k}Y_\ell}$, as defined in Section 2.1.

Let $M \ge d$ and consider the set $V = \{v(1), v(2), \cdots, v(M)\}$ of M unit vectors in $\mathbb{C}^d$. The matrix

$$F = \frac{1}{\sqrt{d}}\begin{bmatrix} \overline{X[1]} & \overline{X[2]} & \cdots & \overline{X[d]} \\ \overline{X[2]} & \overline{X[3]} & \cdots & \overline{X[d+1]} \\ \vdots & \vdots & \cdots & \vdots \\ \overline{X[M]} & \overline{X[M+1]} & \cdots & \overline{X[M+d-1]} \end{bmatrix}$$

is the matrix of the analysis operator corresponding to V. The frame operator of V is $\mathcal{F} = F^*F$, i.e.,

$$\mathcal{F} = \frac{1}{d}\begin{bmatrix} X[1] & X[2] & \cdots & X[M] \\ X[2] & X[3] & \cdots & X[M+1] \\ \vdots & \vdots & \cdots & \vdots \\ X[d] & X[d+1] & \cdots & X[M+d-1] \end{bmatrix}\begin{bmatrix} \overline{X[1]} & \overline{X[2]} & \cdots & \overline{X[d]} \\ \overline{X[2]} & \overline{X[3]} & \cdots & \overline{X[d+1]} \\ \vdots & \vdots & \cdots & \vdots \\ \overline{X[M]} & \overline{X[M+1]} & \cdots & \overline{X[M+d-1]} \end{bmatrix}$$

The entries of $\mathcal{F}$ are given by $\mathcal{F}_{m,m} = \frac{M}{d}$ and for $m \neq n,\ m > n,$

$$\mathcal{F}_{m,n} = \frac{1}{d}\begin{pmatrix} X[m] & X[m+1] & \cdots & X[M+m-1] \end{pmatrix}\begin{pmatrix} \overline{X[n]} \\ \overline{X[n+1]} \\ \vdots \\ \overline{X[M+n-1]} \end{pmatrix}$$

$$= \frac{M}{d}\frac{1}{M}\sum_{\ell=0}^{M-1} X[m+\ell]\overline{X[n+\ell]} = \frac{M}{d}\left(\frac{1}{M}\sum_{\ell=0}^{M-1} e^{\frac{2\pi}{\varepsilon}i(Y_{-m-\ell}+\cdots+Y_{-n-\ell-1}+Y_{n+\ell+1}+\cdots+Y_{m+\ell})}\right).$$

Note that since $\mathcal{F}$ is self-adjoint, $\mathcal{F}_{m,n} = \overline{\mathcal{F}}_{n,m}$. It is desired that V emulates a tight frame, i.e, $\mathcal{F}$ is close to a constant times the identity, in this case, $\frac{M}{d}$ times the identity. Alternatively, it is desirable that the eigenvalues of $\mathcal{F}$ are all close to each other and close to $\frac{M}{d}$. In this case, due to the stochastic nature of the frame operator, one studies the expectation of the eigenvalues of $\mathcal{F}$.

4.1. Frames in $\mathbb{C}^2$

This section discusses the construction of sets of vectors in $\mathbb{C}^2$ as given by (8). The frame properties of such sets are analyzed. In fact, it is shown that the expectation of the eigenvalues of the frame operator are close to each other, the closeness increasing with the size of the set. The bounds on the probability of deviation of the eigenvalues from the expected value is also derived. The related inequalities arise from an application of Theorem 4.1 [22] below.

Theorem 4.1. (Azuma's Inequality) Suppose that $\{X_k : k = 0,1,2,\cdots\}$ is a martingale and

$$|X_k - X_{k-1}| < c_k,$$

almost surely. Then for all positive integers N and all positive reals t,

$$P(|X_N - X_0| \geq t) \leq 2e^{\left(-t^2 \Big/ \left(2\sum_{k=1}^{N} c_k^2\right)\right)}.$$

Consider $M \geq 3$ vectors in $\mathbb{C}^2$ i.e., $d = 2$ in (8). Then $v : \mathbb{Z} \to \mathbb{C}^2$ and

$$v(1) = \frac{1}{\sqrt{2}}\begin{pmatrix} X[1] \\ X[2] \end{pmatrix}, v(2) = \frac{1}{\sqrt{2}}\begin{pmatrix} X[2] \\ X[3] \end{pmatrix}, \cdots, v(M) = \frac{1}{\sqrt{2}}\begin{pmatrix} X[M] \\ X[M+1] \end{pmatrix}. \tag{9}$$

Considering the set $V=\{v(1),v(2),\cdots,v(M)\}$, the frame operator of V is

$$\mathcal{F}=\frac{1}{2}\begin{bmatrix} X[1] & X[2] & X[3] & \cdots & X[M] \\ X[2] & X[3] & X[4] & \cdots & X[M+1]\end{bmatrix}\begin{bmatrix} \overline{X[1]} & \overline{X[2]} \\ \overline{X[2]} & \overline{X[3]} \\ \overline{X[3]} & \overline{X[4]} \\ \vdots & \vdots \\ \overline{X[M]} & \overline{X[M+1]}\end{bmatrix}$$

$$\text{or, } \mathcal{F}=\frac{M}{2}\begin{bmatrix} 1 & \frac{1}{M}\sum_{m=1}^{M} X[m]\overline{X[m+1]} \\ \frac{1}{M}\sum_{m=1}^{M}\overline{X[m]}X[m+1] & 1\end{bmatrix}. \quad (10)$$

Theorem 4.2. 1) Consider the set $V=\{v(1),v(2),\cdots,v(M)\}\in\mathbb{C}^2$, $M\geq 3$, where the vectors $v(n)$ are given by (9). The minimum eigenvalue, $\lambda_{\min}(\mathcal{F})$, and the maximum eigenvalue, $\lambda_{\max}(\mathcal{F})$, of the frame operator of V satisfy

$$\frac{M}{2}(1-\delta)\leq E(\lambda_{\min}(\mathcal{F}))\leq E(\lambda_{\max}(\mathcal{F}))\leq\frac{M}{2}(1+\delta) \quad (11)$$

where $\delta=\sqrt{\frac{1}{M}+\frac{M-1}{M}e^{-2\sigma^2\left(\frac{2\pi}{\varepsilon}\right)^2}}$.

2) The deviation of the minimum and maximum eigenvalue of $\mathcal{F}$ from their expected value is given, for all positive reals r, by

$$P\left(\left|\lambda_{\min}(\mathcal{F})-E(\lambda_{\min}(\mathcal{F}))\right|>r\right)\leq 2e^{-\frac{4r^2}{8M^3}},$$

$$P\left(\left|\lambda_{\max}(\mathcal{F})-E(\lambda_{\max}(\mathcal{F}))\right|>r\right)\leq 2e^{-\frac{4r^2}{8M^3}}.$$

Proof. 1) The frame operator of

$V=\{v(1),v(2),\cdots,v(M)\}$ is given in (10). The eigenvalues of $\frac{2}{M}\mathcal{F}$ are $\lambda_1=1-|\alpha|$ and $\lambda_2=1+|\alpha|$ where

$$\alpha = \frac{1}{M}\sum_{m=1}^{M} X[m]\overline{X[m+1]}.$$

Let

$$\begin{aligned}
\gamma_1 &= X[1]\overline{X[2]} = \mathrm{e}^{-\frac{2\pi}{\varepsilon}i(Y_{-2}+Y_2)},\\
\gamma_2 &= X[2]\overline{X[3]} = \mathrm{e}^{-\frac{2\pi}{\varepsilon}i(Y_{-3}+Y_3)},\\
&\vdots\\
\gamma_M &= X[M]\overline{X[M+1]} = \mathrm{e}^{-\frac{2\pi}{\varepsilon}i\left(Y_{-(M+1)}+Y_{(M+1)}\right)},
\end{aligned}$$

so that

$$\alpha = \frac{\gamma_1+\gamma_2+\cdots+\gamma_M}{M}.$$

Note that for $m \neq n, \gamma_m$ and γ_n are independent and so $E\left(\gamma_m\overline{\gamma_n}\right) = E(\gamma_m)E\left(\overline{\gamma_n}\right)$. Also, since the Y_ℓ s are i.i.d. and the characteristic function of the Y_ℓ s is symmetric,

$\forall\, 1 \leq m \leq M,$

$$\begin{aligned}
E(\gamma_m) &= E\left(\mathrm{e}^{\frac{2\pi}{\varepsilon}i\left(Y_{-(m+1)}+Y_{(m+1)}\right)}\right) = \left\{E\left(\mathrm{e}^{\frac{2\pi}{\varepsilon}iY_1}\right)\right\}^2\\
&= \mathrm{e}^{-\sigma^2\left(\frac{2\pi}{\varepsilon}\right)^2} = E\left(\overline{\gamma_m}\right)
\end{aligned}$$

and therefore

$$E\left(\gamma_m\overline{\gamma_n}\right) = \mathrm{e}^{-2\sigma^2\left(\frac{2\pi}{\varepsilon}\right)^2}.$$

Thus

$$
\begin{aligned}
E\left(|\alpha|^2\right) &= E(\alpha\bar{\alpha}) \\
&= \frac{1}{M^2} E\left((\gamma_1+\gamma_2+\cdots+\gamma_M)\left(\overline{\gamma_1}+\overline{\gamma_2}+\cdots+\overline{\gamma_M}\right)\right) \\
&= \frac{1}{M^2} E\left(|\gamma_1|^2+|\gamma_2|^2+\cdots+|\gamma_M|^2+\sum_{m\neq n}\gamma_m\overline{\gamma_n}\right) \\
&= \frac{1}{M}+\frac{1}{M^2} E\left(\sum_{m\neq n}\gamma_m\overline{\gamma_n}\right) = \frac{1}{M}+\frac{1}{M^2}\sum_{m\neq n} E\left(\gamma_m\overline{\gamma_n}\right) \\
&= \frac{1}{M}+\frac{M-1}{M}\mathrm{e}^{-2\sigma^2\left(\frac{2\pi}{\varepsilon}\right)^2}.
\end{aligned}
$$

The above estimate on $E\left(|\alpha|^2\right)$ implies that

$$
E(|\alpha|) \leq \sqrt{E\left(|\alpha|^2\right)} = \sqrt{\frac{1}{M}+\frac{M-1}{M}\mathrm{e}^{-2\sigma^2\left(\frac{2\pi}{\varepsilon}\right)^2}}. \tag{12}
$$

Since $E(\lambda_1) = 1 - E(|\alpha|)$ and $E(\lambda_2) = 1 + E(|\alpha|)$, (12) implies

$$
\begin{aligned}
1-\sqrt{\frac{1}{M}+\frac{M-1}{M}\mathrm{e}^{-2\sigma^2\left(\frac{2\pi}{\varepsilon}\right)^2}} &\leq E(\lambda_1) \leq E(\lambda_2) \\
&\leq 1+\sqrt{\frac{1}{M}+\frac{M-1}{M}\mathrm{e}^{-2\sigma^2\left(\frac{2\pi}{\varepsilon}\right)^2}}.
\end{aligned}
$$

Noting that $\lambda_{\min}(\mathcal{F}) = \frac{M}{2}\lambda_1$ and $\lambda_{\max}(\mathcal{F}) = \frac{M}{2}\lambda_2$, one finally gets, after setting $\delta = \sqrt{\frac{1}{M}+\frac{M-1}{M}\mathrm{e}^{-2\sigma^2\left(\frac{2\pi}{\varepsilon}\right)^2}}$,

$$
\frac{M}{2}(1-\delta) \leq E\left(\lambda_{\min}(\mathcal{F})\right) \leq E\left(\lambda_{\max}(\mathcal{F})\right) \leq \frac{M}{2}(1+\delta).
$$

2) To prove 2) we use the Doob martingale and Azuma's inequality [22]. For $n = 2,\cdots,M+1$, let $Z_{n-1} = Y_{-n} + Y_n$. Here the Doob martingale is the sequence $\{U_0, U_1, \cdots, U_{M-1}\}$ where

$$\text{for } k = 1, \cdots, M-1,$$
$$U_k = E\left(\frac{1}{M} \left| \sum_{j=1}^{M} e^{\frac{2\pi}{\varepsilon} iZ_j} \right| \middle| Z_1, Z_2, \cdots, Z_k \right)$$

and

$$U_0 = E\left(\frac{1}{M} \left| \sum_{j=1}^{M} e^{\frac{2\pi}{\varepsilon} iZ_j} \right| \right).$$

Note that $U_0 = E(|\alpha|)$ and $U_{M-1} = |\alpha|$. Also,

$$|U_k - U_{k-1}| \le |U_k| + |U_{k-1}| \le 2.$$

So by Azuma's Inequality (see Theorem 4.1)

$$P(|U_{M-1} - U_0| \ge r) = P(||\alpha| - E(|\alpha|)| \ge r)$$
$$\le 2e^{-r^2 / \left(2\sum_{k=1}^{M} 2^2\right)} = 2e^{-r^2/8M}.$$

Since $|\lambda_1 - E(\lambda_1)| = |\lambda_2 - E(\lambda_2)| = ||\alpha| - E(|\alpha|)|$, this means

$$P(|\lambda_1 - E(\lambda_1)| > r) \le 2e^{-r^2/8M}$$

and

$$P(|\lambda_2 - E(\lambda_2)| > r) \le 2e^{-r^2/8M}$$

Going back to the actual frame operator $\mathcal{F}$, whose eigenvalues are $\frac{M}{2}\lambda_1$ and $\frac{M}{2}\lambda_2$, the following estimates hold.

$$P(|\lambda_{\max}(\mathcal{F}) - E(\lambda_{\max}(\mathcal{F}))| > r)$$
$$= P\left(\frac{M}{2} ||\alpha| - E(|\alpha|)| > r \right)$$
$$= P\left(||\alpha| - E(|\alpha|)| > \frac{2}{M} r \right) \le 2e^{-\frac{4r^2}{8M^3}}$$

and

$$
\begin{aligned}
&P\left(\left|\lambda_{\min}(\mathcal{F}) - E\left(\lambda_{\min}(\mathcal{F})\right)\right| > r\right) \\
&= P\left(\frac{M}{2}\left|\|\alpha\| - E(\|\alpha\|)\right| > r\right) \\
&= P\left(\left|\|\alpha\| - E(\|\alpha\|)\right| > \frac{2}{M} r\right) \le 2\mathrm{e}^{-\frac{4r^2}{8M^3}}. \quad \square
\end{aligned}
$$

Corollary 4.3. The eigenvalues of the frame operator considered in Theorem 4.2 satisfy, for all positive reals r,

$$P\left(\lambda_{\min}(\mathcal{F}) < \frac{M}{2}(1-\delta) - r\right) \le \mathrm{e}^{-4r^2/8M^3},$$

$$P\left(\lambda_{\max}(\mathcal{F}) > \frac{M}{2}(1+\delta) + r\right) \le \mathrm{e}^{-4r^2/8M^3},$$

where $\delta = \sqrt{\frac{1}{M} + \frac{M-1}{M}\mathrm{e}^{-2\sigma^2\left(\frac{2\pi}{\varepsilon}\right)^2}}$.

Proof. Due to part 1) of Theorem 4.2

$$\lambda_{\min}(\mathcal{F}) < \frac{M}{2}(1-\delta) - r \Rightarrow \lambda_{\min}(\mathcal{F}) < E\left(\lambda_{\min}(\mathcal{F})\right) - r.$$

This implies, as a consequence of part 2) of Theorem 4.2, that

$$
\begin{aligned}
&P\left(\lambda_{\min}(\mathcal{F}) < \frac{M}{2}(1-\delta) - r\right) \\
&\le P\left(\lambda_{\min}(\mathcal{F}) < E\left(\lambda_{\min}(\mathcal{F})\right) - r\right) \le \mathrm{e}^{-4r^2/8M^3}.
\end{aligned}
$$

In a similar way, from part 1) of Theorem 4.2,

$$\lambda_{\max}(\mathcal{F}) > \frac{M}{2}(1+\delta) + r \Rightarrow \lambda_{\max}(\mathcal{F}) > E\left(\lambda_{\max}(\mathcal{F})\right) + r$$

which implies, as a consequence of part 2) of Theorem 4.2, that

$$
\begin{aligned}
&P\left(\lambda_{\max}(\mathcal{F}) > \frac{M}{2}(1+\delta) + r\right) \\
&\le P\left(\lambda_{\max}(\mathcal{F}) > E\left(\lambda_{\max}(\mathcal{F})\right) + r\right) \le \mathrm{e}^{-4r^2/8M^3}. \quad \square
\end{aligned}
$$

Remark 4.4. In Theorem 4.2, as M tends to infinity, the value of δ in (11) can be made arbitrarily small based on the choice of ε. This in turn implies that the two eigenvalues can be made arbitrarily close to each other, with ε. On the other hand, for a fixed M, as ε tends to zero, (11) becomes

$$\frac{M}{2}\left(1-\sqrt{\frac{1}{M}}\right) \le E\left(\lambda_{\min}(\mathcal{F})\right)$$
$$\le E\left(\lambda_{\max}(\mathcal{F})\right) \le \frac{M}{2}\left(1+\sqrt{\frac{1}{M}}\right).$$

4.2. Frames in $\mathbb{C}^d$; $d > 2$

For general d and M, in order to use existing results on the concentration of eigenvalues of random matrices [23, 24], a slightly different construction of the frame needs to be considered. Let $\{Y_{mn}\}_{m,n \subset \mathbb{Z}}$ be i.i.d. random variables following a Gaussian distribution with mean zero and variance σ^2. It can be shown that

$$E\left(e^{\frac{2\pi}{\varepsilon} iY_{mn}}\right) = e^{-\frac{\sigma^2}{2}\left(\frac{2\pi}{\varepsilon}\right)^2}$$

and the variance

$$V\left(e^{\frac{2\pi}{\varepsilon} iY_{mn}}\right) = 1 - e^{-\sigma^2\left(\frac{2\pi}{\varepsilon}\right)^2}.$$

One can define the following two dimensional sequence. For $m, n \in \mathbb{Z}$,

$$X_{mn} = e^{\frac{2\pi}{\varepsilon} iY_{mn}} - e^{-\frac{\sigma^2}{2}\left(\frac{2\pi}{\varepsilon}\right)^2}.$$

Consider the mapping $v : \mathbb{Z} \to \mathbb{C}^d$ given by

$$v(\ell)=\frac{1}{\sqrt{d}}\begin{pmatrix} X_{1\ell} \\ X_{2\ell} \\ \vdots \\ X_{d\ell} \end{pmatrix}. \quad (13)$$

As before, let $M \geq d$ and consider the set of M unit vectors $V=\{v(1),v(2),\cdots,v(M)\}$ in $\mathbb{C}^d$. The frame operator of this set is

$$\mathcal{F}=\frac{1}{d}\begin{bmatrix} X_{11} & X_{12} & \cdots & X_{1M} \\ X_{21} & X_{22} & \cdots & X_{2M} \\ \vdots & \vdots & \cdots & \vdots \\ X_{d1} & X_{d2} & \cdots & X_{dM} \end{bmatrix}\begin{bmatrix} \overline{X}_{11} & \overline{X}_{21} & \cdots & \overline{X}_{d1} \\ \overline{X}_{12} & \overline{X}_{22} & \cdots & \overline{X}_{d2} \\ \vdots & \vdots & \cdots & \vdots \\ \overline{X}_{1M} & \overline{X}_{2M} & \cdots & \overline{X}_{dM} \end{bmatrix}.$$

Let

$$A=\frac{1}{\sqrt{d}}\begin{bmatrix} X_{11} & X_{12} & \cdots & X_{1M} \\ X_{21} & X_{22} & \cdots & X_{2M} \\ \vdots & \vdots & \cdots & \vdots \\ X_{d1} & X_{d2} & \cdots & X_{dM} \end{bmatrix} \quad (14)$$

so that $\mathcal{F}=AA^*$. The matrix A has entries with mean zero and variance $\hat{\sigma}^2=\frac{1}{d}\left(1-e^{-\sigma^2\left(\frac{2\pi}{\varepsilon}\right)^2}\right)$. According to results in [23], if $\frac{d}{M}\to c$ as $d,\ M\to\infty$, then the smallest and largest eigenvalues of $\mathcal{F}$ converge almost surely to $\hat{\sigma}^2\left(1-\sqrt{c}\right)^2$ and $\hat{\sigma}^2\left(1+\sqrt{c}\right)^2$, respectively.

Theorem 4.5. Let $s_1(A)\leq s_2(A)\leq\cdots\leq s_d(A)$ be the singular values of the matrix A given by (14). Then the following hold.

1) Given ε_0, there is a large enough d such that

$$P\left(s_d(A)\geq\hat{\sigma}\left(1+\sqrt{\frac{d}{M}}\right)+\varepsilon_0+r\right)\leq 2e^{-r^2d/16}. \quad (15)$$

2) $P\left(s_1(A) \le c_1\right) \le e^{-c_2 M}$, (16)

where c_1 and c_2 are universal positive constants.

Proof. Let s_d be the mapping that associates to a matrix A it largest singular value. Equip $\mathbb{C}^{dM}$ with the Frobenius norm

$$\|A\|^2 := Tr\left(AA^*\right) = \sum_{m,n} |A_{mn}|^2 .$$

Then the mapping s_d is convex and 1-Lipschitz in the sense that

$$\left|s_d(A) - s_d(A')\right| \le \|A - A'\|$$

for all pairs (A, A') of d by M matrices [24].

We think of A as a random vector in $\mathbb{R}^{2dM}$. The real and imaginary parts of the entries of $\frac{1}{\sqrt{d}}A$ are supported in $\left[-\frac{1}{\sqrt{d}}, \frac{1}{\sqrt{d}}\right]$. Let P be a product measure on $\left[-\frac{1}{\sqrt{d}}, \frac{1}{\sqrt{d}}\right]^{2dM}$. Then as a consequence of the concentration inequality (Corollary 4.10, [24]) we have

$$P\left(\left|s_d(A) - m(s_d)\right| \ge r\right) \le 4e^{-r^2 d/16}$$

where $m(s_d)$ is the median of $s_d(A)$. It is known that the minimum and maximum singular values of A converge almost surely to $\hat{\sigma}\left(1-\sqrt{c}\right)$ and $\hat{\sigma}\left(1+\sqrt{c}\right)$, respectively, as d, M tend to infinity and $\frac{d}{M} \to c$. As a consequence, for each ε_0 and M sufficiently large, one can show that the medians belong to the fixed interval

$$\left[\hat{\sigma}\left(1-\sqrt{\frac{d}{M}}\right) - \varepsilon_0, \hat{\sigma}\left(1+\sqrt{\frac{d}{M}}\right) + \varepsilon_0\right]$$

which gives

$$P\left(s_d(A) \ge \hat{\sigma}\left(1+\sqrt{\frac{d}{M}}\right) + \varepsilon_0 + r\right) \le 2e^{-r^2 d/16}.$$

For the smallest singular value we cannot use the concentration inequality as used for s_d since the smallest singular value is not convex. However, following results in [25] (Theorem 3.1) that have been used in [26] in a similar situation as here, one can say that whenever $M > (1+\delta)d$, where δ is greater than a small constant,

$P(s_1(A) \le c_1) \le e^{-c_2 M}$ where c_1 and c_2 are positive universal constants. □

Remark 4.6. Note that the square of the singular values of A are the eigenvalues of $\mathcal{F}$ and so the estimates given in (15) and (16) give insight into the corresponding deviation of the eigenvalues of the frame operator $\mathcal{F}$.

Remark 4.7. (Connection to compressed sensing) The theory of compressed sensing [27-29] states that it is possible to recover a sparse signal from a small number of measurements. A signal $x \in \mathbb{C}^M$ is k-sparse in a basis $\Psi = \{\psi_j\}_{j=1}^{M}$ if x is a weighted superposition of at most k elements of Ψ. Compressed sensing broadly refers to the inverse problem of reconstructing such a signal x from linear measurements $\{y_\ell = \langle x, \phi_\ell \rangle | \ell = 1, \cdots, d\}$ with $d < M$, ideally with $d \ll M$. In the general setting, one has $\Phi^* x = y$, where Φ is a $M \times d$ sensing matrix having the measurement vectors ϕ_ℓ as its columns, x is a length-M signal and y is a length-d measurement.

The standard compressed sensing technique guarantees exact recovery of the original signal with very high probability if the sensing matrix satisfies the Restricted Isometry Property (RIP). This means that for a fixed k, there exists a small number δ_k, such that

$$(1-\delta_k)\|x\|_{\ell_2}^2 \le \|\Phi^* x\|_{\ell_2}^2 \le (1+\delta_k)\|x\|_{\ell_2}^2,$$

for any k-sparse signal x. By imitating the work done in [26] (Lemmas 4.1 and 4.2), it can be shown, due to Theorem 4.5, that matrices A of the type given in (14) satisfy the RIP condition and can therefore be used as measurement matrices in compressed sensing. These matrices are different from the traditional random matrices used in compressed sensing in that their entries are complexvalued and unimodular instead of being real-valued and not unimodular.

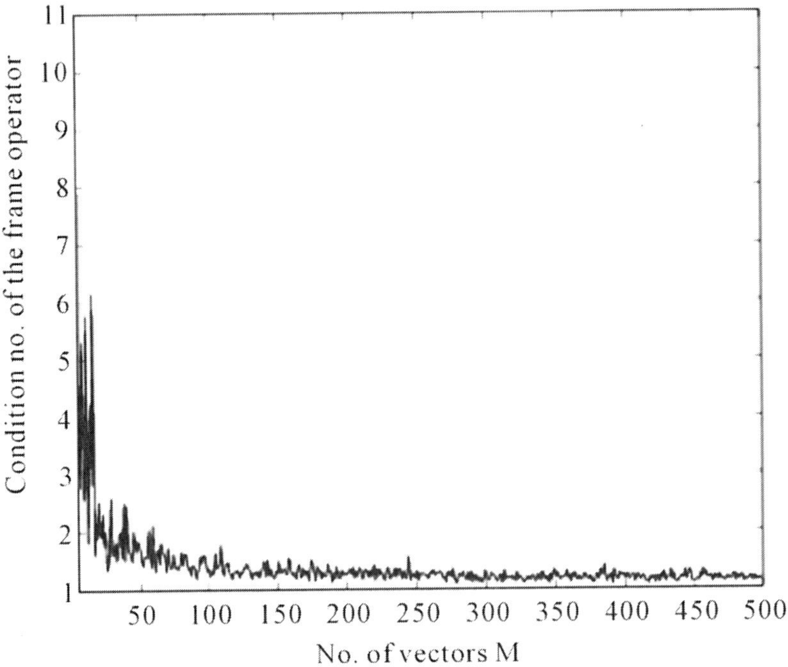

Figure 1. Behavior of the condition number of the frame operator with increasing size of the frame; ε = 0.0001, d = 3, σ = 1.

Example 4.8. This example illustrates the ideas in this subsection. First consider M = 5 and d = 3 so that there are 5 vectors in $\mathbb{C}^3$. Taking from a normal distribution with mean 0 and variance $\sigma = 1$, a realization of the matrix $\left[Y_{mn}\right]_{1 \le m \le 3, 1 \le n \le 5}$ is

$$\begin{bmatrix} -0.0353 & 0.5004 & -0.6299 & -0.1472 & 0.4003 \\ -0.4804 & -0.9344 & 0.4220 & -0.9509 & 0.2783 \\ -0.8609 & -0.4822 & -0.4680 & -0.9509 & 1.2284 \end{bmatrix}.$$

Then taking $\varepsilon = 0.001$, $A = \frac{1}{\sqrt{3}}\left[e^{\frac{2\pi}{\varepsilon} i Y_{mn}}\right]$ is

$$A = \frac{1}{\sqrt{3}}\begin{bmatrix} -0.27-0.96i & -0.89+0.44i & 0.85+0.52i & 0.33-0.94i & -0.24+0.97i \\ -0.92-0.39i & -0.89-0.46i & 0.99+0.05i & 0.93+0.37i & -0.47+0.88i \\ 0.74+0.68i & 0.16-0.99i & 0.99+0.09i & -0.30-0.95i & -0.74+0.67i \end{bmatrix}.$$

The condition number, ratio of the maximum and minimum eigenvalues, of $\mathcal{F} = 4.8667$. As the number of vectors M is increased, the condition number gets closer to 1. Figure 1 shows the behavior of the condition number with the increase in the number of vectors.

5. CONCLUSION

The construction of discrete unimodular stochastic waveforms with arbitrarily small expected autocorrelation has been proposed. This is motivated by the usefulness of such waveforms in the areas of radar and communications. The family of random variables that can be used for this purpose has been characterized. Such construction been done in one dimension and generalized to higher dimensions. Further, such waveforms have been used to construct frames in $\mathbb{C}^d$ and the frame properties of such frames have been studied. Using Brownian motion, this idea is also extended to the construction of continuous unimodular stochastic waveforms whose autocorrelation can be made arbitrarily small in expectation.

ACKNOWLEDGEMENTS

The author wishes to acknowledge support from AFOSR Grant No. FA9550-10-1-0441 for conducting this research. The author is also grateful to Frank Gao and Ross Richardson for their generous help with probability theory.

REFERENCES

1. L. Auslander and P. E. Barbano, "Communication Codes and Bernoulli Transformations," Applied and Computational Harmonic Analysis, Vol. 5, No. 2, 1998, pp. 109- 128.

2. J. J. Benedetto and J. J. Donatelli, "Ambiguity Function and Frame Theoretic Properties of Periodic Zero Autocorrelation Waveforms," IEEE Journal of Selected Topics in Signal Processing, Vol. 1, 2007, pp. 6-20.

3. T. Helleseth and P. V. Kumar, "Sequences with Low Correlation," Handbook of Coding Theory, North-Holland, Amsterdam, 1998, pp. 1765-1853.

4. N. Levanon and E. Mozeson, "Radar Signals," Wiley Interscience, New York, 2004.

5. M. L. Long, "Radar Reflectivity of Land and Sea," Artech House, 2001.

6. W. H. Mow, "A New Unified Construction of Perfect Root-of-Unity Sequences," Proceedings of IEEE 4th International Symposium on Spread Spectrum Techniques and Applications, Sun City, 22-25 September 1996, pp. 955-959.

7. F. E. Nathanson, "Radar Design Principles: Signal Processing and the Environment," SciTech Publishing Inc., Mendham, 1999.

8. G. W. Stimson, "Introduction to Airborne Radar," SciTech Publishing Inc., Mendham, 1998.

9. S. Ulukus and R. D. Yates, "Iterative Construction of Optimum Signature Sequence Sets in Synchronous CDMA Systems," IEEE Transactions on Information Theory, Vol. 47, No. 5, 2001, pp. 1989-1998.

10. S. Verdú, "Multiuser Detection," Cambridge University Press, Cambridge, 1998.

11. J. J. Benedetto and S. Datta, "Construction of Infinite Unimodular Sequences with Zero Autocorrelation," Advances in Computational Mathematics, Vol. 32, No. 2, 2010, pp. 191-207.

12. D. Cochran, "Waveform-Agile Sensing: Opportunities and Challenges," IEEE International Conference on Acoustics, Speech, and Signal Processing, Pennsylvania, 18-23 March 2005, pp. 877-880.

13. M. R. Bell, "Information Theory and Radar Waveform Design," IEEE Transactions on Information Theory, Vol. 39, No. 5, 1993, pp. 1578-1597.

14. S. P. Sira, Y. Li, A. Papandreou-Suppappola, D. Morrell, D. Cochran and M. Rangaswamy, "Waveform-Agile Sensing for Tracking," IEEE of Signal Processing Magazine, Vol. 26, No. 1, 2009, pp. 53-64.

15. H. Boche and S. Stanczak, "Estimation of Deviations between the Aperiodic and Periodic Correlation Functions of Polyphase Sequences in Vicinity of the Zero Shift," IEEE 6th International Symposium on Spread Spectrum Techniques and Applications, Parsippany, 6-8 September 2000, pp. 283-287.

16. R. Narayanan, "Through Wall Radar Imaging Using UWB Noise Waveforms," IEEE International Conference on Acoustics, Speech and Signal Processing, Beijing, 31 March-4 April 2008, pp. 5185-5188.

17. R. M. Narayanan, Y. Xu, P. D. Hoffmeyer and J. O. Curtis, "Design, Performance, and Applications of a Coherent Ultra-Wideband Random Noise Radar," Optical Engineering, Vol. 37, No. 6, 1998, pp. 1855-1869.

18. D. Bell and R. Narayanan, "Theoretical Aspects of Radar Imaging Using Stochastic Waveforms," IEEE Transactions on Signal Processing, Vol. 49, No. 2, 2001, pp. 394-400.

19. F. Karr, "Probability of Springer Texts in Statistics," Springer-Verlag, New York, 1993.

20. O. Christensen, "An Introduction to Frames and Riesz Bases," Birkhäuser, 2003.

21. Daubechies, "Ten Lectures on Wavelets," CBMS-NSF Regional Conference Series in Applied Mathematics, 1992. doi:10.1137/1.9781611970104

22. W. Hoeffding, "Probability Inequalities for Sums of Bounded Random Variables," Journal of American Statistical Association, Vol. 58, No. 301, 1963, pp. 13-30.

23. Z. D. Bai, "Methodologies in Spectral Analysis of LargeDimensional Random Matrices: A Review," Statistica Sinica, Vol. 9, No. 3, 1999, pp. 611-677.

24. M. Ledoux, "The Concentration of Measure Phenomenon," Mathematical Surveys and Monographs, American Mathematical Society, 2001.

25. E. Litvak, A. Pajor, M. Rudelson and N. TomczakJaegermann, "Smallest Singular Value of Random Matrices and Geometry of Random Polytopes," Advances in Mathematics, Vol. 195, No. 2, 2005, pp. 491-523.

26. E. Candès and T. Tao, "Near Optimal Signal Recovery from Random Projections: Universal Encoding Strategies?" IEEE Transactions on Information Theory, Vol. 52, No. 12, 2006, pp. 5406-5425.

27. E. Candès, "Compressive Sampling," Proceedings of the International Congress of Mathematicians, Madrid, 22-30 August 2006, pp. 1433-1452.

28. E. Candès, J. Romberg and T. Tao, "Robust Uncertainty Principles: Exact Signal Reconstruction from Highly Incomplete Frequency Information," IEEE Transactions on Information Theory, Vol. 52, No. 2, 2006, pp. 489-509.

29. D. L. Donoho, "Compressed Sensing," IEEE Transactions on Information Theory, Vol. 52, No. 4, 2006, pp. 1289- 1306.

NOTES

*This work was supported by AFOSR Grant No. FA9550-10-1-0441.

Index